Methane: Planning a Digester

C.T.T. Series No. 2.

Methane:
Planning a Digester

by: Peter – John Meynell

Prism Press

Published 1976 in Great Britain by

Prism Press
Stable Court
Chalmington
Dorchester
Dorset

© P.J. Meynell 1976
Drawings by Keith Bennett.

ISBN 0 904727 11 4 Hardback
ISBN 0 904727 12 2 Paperback

Printed by

Unwin Brothers Limited
The Gresham Press
Old Woking
Surrey

Acknowledgments

I must thank all the people and organisations who have been so helpful in providing me with information about methane and about their efforts in trying to produce it efficiently (see Chapter 9). Of those mentioned I would particularly like to thank the people who gave me diagrams and photographs with which to illustrate this book, namely Dr. Hobson and Messrs. Horton, Thompson, Mitchell, Rippon and Knox.

Above all I must thank Keith Bennett for his excellent diagrams which help so much in the understanding of the process. Others without whom the book would never have been possible were Stewart Owen, Simon and Pippa Gibbs and Carol Frayne. In this respect most thanks must go to Judith Meynell who kept my English up to scratch, who read and simplified the contents and who did the bulk of the typing.

I would also like to thank Hugh Sharman and all those at CTT who helped me, stimulated my initial interest in methane and who provided the contact with Prism Press.

Contents

Introduction

by HUGH SHARMAN

Throughout the previous 150 years, and in particular during the last 20, there has been a dramatic increase in man's use of stored energy, for industrial manufacturing processes, transportation and personal comfort. Most of our energy has been provided in the form of ancient stores such as coal, oil and natural gas. More recent sources, such as wood, are already severely depleted with forest disappearing to desert at an alarming rate. The industrial epoch has suffered a number of energy crises, mostly local, minor and temporary. Many ingenious solutions were discovered. For instance, during the last war Germany made liquid fuel from coal and British cars were driven by coal gas stored on the roof. During the late 1940's the buses on which I travelled in China were powered by the alcohols driven off wood by charcoal burners.

The present energy crisis however, which began in 1969, is of a very different order. During the 1940's, 50's and 60's improved methods of locating, extracting and refining oil products together with the West's post-colonial grip on the world economy, enabled the cost of oil-based energy to fall in relation to almost all other major commodities. The extra industrial productivity which this was able to generate led to a spectacular increase in demand. Also by 1969 Geophysicists were predicting that if consumption continued to increase reserves might be exhausted by the end of the century. This then was the first truly global 'energy crisis'. As with all crises a huge and varied response was evoked, including proposals for an 'alternative technology'. Many of the suggestions were particularly stupid and counter-productive. The 'high technologists' advocated, and indeed are still advocating, high technology, high energy solutions which might even result in a net energy loss. For instance, the use of nuclear fission, nuclear fusion and photo-electric satellites. Almost no less fatuous are many of the solutions which have been put forward by the 'alternative technologists'. The idea that late 20th century comfort and convenience can be sustained by whirling windmills and solar collectors has already been pedalled by quite a number of charlatans. High on the list of alternative technologies that have been oversold is methane gas. For this reason Peter Meynell's latest contribution to the discussion is very welcome.

Methane Gas (CH_4) occurs widely in nature and is called natural gas. It is the simplest of all naturally occurring hydrocarbons. Here in Britain, we have been using methane as a prime energy source for cooking, domestic heating, steam raising etc. since the discovery and exploitation of gas in the southern area of the North Sea. These huge reserves of gas were already declining by 1972 and the British Gas Corporation is relying absolutely on the discovery and exploitation of oil associated gas and other gas fuels in the northern area of the North Sea. This is bound to result in a very much more expensive product and will barely extend the life of British fossil gas by more than a decade or so.

Significantly, 1972 also marked the publication of a number of unrealistic books and articles about methane. Television programmes were broadcast describing the activities of an energetic Devon farmer who drove his car on 'chicken shit'. Numerous home-made digesters were built with the object of driving cars, warming houses etc. etc. The reality turns out to be much less exciting.

However, there can be no doubt that anaerobic digestion has a very real use in Britain during the years that lie ahead. In India, for example, the soil has been subjected to a spiral of deprivation because the peasants burn the animal dung as fuel. The Gobar Gas Institute has designed small digesters that can provide a family with five cows enough gas for cooking food and boiling water. The digested refuse, high in nitrogen and organic matter, is a first class soil conditioner and fertiliser. The fuel itself burns cleanly and produces no unpleasant by-products. Here in lush Britain, with one of the most equitable climates in the world, we have been guilty of a neglect less excusable than that of the Indian peasant. Many inches of valuable topsoil have disappeared from our prairie wheatlands. Organic filling, so necessary for good soil structure, has likewise disappeared after successive seasons of planting without compost. Our agriculture has been depending more and more upon imported inputs of chemical fertilisers while the natural waste products of the farm have been dumped wherever local authorities could make space for them, both unused and unuseable. Thus it is in the fields of agriculture and pollution control that anaerobic digestion will have the most beneficial impact.

It is little realised by the population at large how little energy is generated by the anaerobic digestion of sewage. 1000 pigs produce only 10kW of gas which in turn could only convert into

3kW of electricity. At least 6kW or so is usually necessary to keep a generator warm enough to make an anaerobic digester work therefore the net energy useable is only 3kW-4kW. But it is surprising what can be done with even small amounts of gas. The fertilising quality of the products are such that at a time when all other forms of energy and fertiliser can be seen to be very finite indeed, and foreign exchange is becoming so expensive, the significance of anaerobic digestion can really be seen.

Peter's book is a masterpiece of cool, intelligent writing. His prose style is simple; neither condescending nor obtruse. His diagrams and illustrations are superb. He has provided a more balanced, scholarly picture of the possibilities of the anaerobic process than any other that I have come across. He has also done much to restore perspectives, so distorted by many of the other books. For this reason I recommend this book to all serious students of energy alternatives as well as potential clients of the process.

1 Introducing Methane

The production of methane gas from organic material is not a new process. Man has known about it ever since Volta discovered methane in marsh gas in the eighteenth century, and ideas and experiments on how the process can be used have been going on for the past hundred years. Interest waned in the middle of this century, when the availability and cheapness of other fuels minimised the need for methane. Lately, with the realisation that other energy sources are limited and expensive, enthusiasm for the process has revived.

Anaerobic Digestion — the process by which methane is produced — has been hailed by some as the answer to all our energy problems: everyone could have his own digester and be self-sufficient. Now that people are learning to live with expensive energy, the time has come to put methane and anaerobic digestion into perspective; to show how and where it is best used and to help people decide whether it is worthwhile exploiting this particular energy source to supplement their existing ones.

There are three natural sources of methane gas, only one of which is commercially exploited. This is the methane found in the various natural gases (e.g. North Sea Gas) which originates from the physical and chemical breakdown of pre-historic plant material. Trapped beneath the earth's crust, it is an important fossil fuel, like coal or oil.

The second natural source can be seen in the strange lights flickering over marshy ground, known as Will-o'-the-Wisps.[1] These are caused by methane escaping from the muds and spontaneously igniting under certain atmospheric conditions. The methane is produced from the decomposition of plants by bacteria in the absence of air.

The third source is from the stomachs of herbivores like cattle and sheep (i.e. ruminants). These animals could not digest the grass that they eat if it were not for anaerobic (in the absence of air) bacteria which break the cellulose in the grass down into molecules which the cow can absorb. One of the

1

by-products of ruminant digestion is methane produced by these bacteria in the stomach, and the animals have to belch in order to rid themselves of it. The condition known as 'bloat' occurs in ruminants when they suddenly change to a very rich diet — e.g. fresh clover — and methane is produced in the stomach faster than the animal can eruct. A balanced diet is needed for a cow just as much as it is needed for man-made digesters, which work on a similar principle to the rumen.

The most widely used man-made digester is that found in sewage works. The process of anaerobic digestion has been exploited ever since men first started treating sewage. The cesspool was the original form of sewage treatment and this led to the development of the septic tank in the middle of the nineteenth century. Both used anaerobic digestion of human excreta, without collecting the methane gas produced. As populations grew improvements were made to these processes and indeed in the 1890's Donald Cameron designed a septic tank for the city of Exeter from which gas was collected and used for street lighting.[2] Later Karl Imhoff in Germany produced a very popular tank (the Imhoff tank) which consisted of two compartments — one where the solids were settled and the second in which these solids were anaerobically digested to produce methane gas.[3]

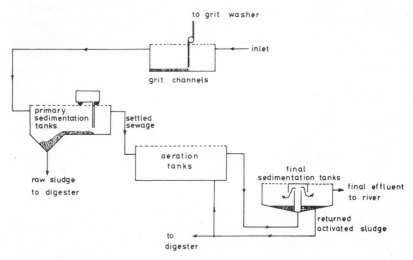

Fig. 1.   Flow diagram of a modern sewage works.

In modern sewage works *(Fig. 1)* the separation of the solids takes place in special settlement tanks, and the water — which makes up 99.9% of sewage — is treated aerobically (in the presence of air) by pumping air through it (the activated-sludge process) or by allowing it to trickle over stones etc. (the percolating filter process). Both methods make use of bacteria to break down the polluting organic matter in the water and both produce a sludge. This secondary sludge and the primary solids are mixed and digested together in special anaerobic digesters, making the sludge easier to dispose of and producing methane as well. One of the earliest separate sewage-sludge digesters was built in Birmingham in 1911.[2] Nowadays most of the larger sewage works have separate digesters and many are entirely self-sufficient in their energy needs (e.g. for pumping and aeration) with the methane produced. However, because most of the applications of the process have been at large sewage works, this has led to the assumption that it is only suitable and economically viable on a large scale. It is only in times of fuel shortage that smaller-scale uses have been tried.

These uses include the treatment of agricultural wastes and more recently industrial ones. During the Second World War, both in England and on the Continent in France and Germany, many farmers built small digesters to produce methane from their farm wastes. The same Karl Imhoff drew up plans for small agricultural power plants[4] to provide electricity and fuel for tractors and farm vehicles. The digesters, in general, were laborious and messy to use, and as other fuel became cheap and readily available most fell into disuse by the end of the 1950's. At some sewage works, during the war, the gas was used for vehicles at the works as a means of saving imported fuel — a matter for national expediency rather than strictly economic reasons.[5]

Abroad, especially in developing countries, the need for a cheap and readily available source of energy is ever present and not merely in times of fuel crises. The longest standing work in the production of methane has been going on in India since the nineteenth century. (The first plant ever recorded was built in a leper colony in Bombay in 1859.)[6] Moreover, the Indian Agricultural Research Institute in 1939 developed the first Gobar (cow dung) gas plant, the success of which led to the formation of the Gobar Gas Institute in the 1950's. This institute was set up to promote the anaerobic digestion of cow dung on a family or village scale so as to provide both a useful fuel for cooking etc. and a valuable fertiliser. Other experiments and farm-scale digesters were developed in Kenya and South Africa in the 1950's. There are reports that the Chinese have been using methane from covered lagoons for communes and factories for several decades,[7] and also that there are small digesters in Taiwan and Sarawak.

In the West general interest in methane digesters revived with the growing awareness of environmental pollution in the late

1960's, which culminated in the energy crisis of 1973. At that time the process was thought to be a universal prescription for 'free' renewable energy which would save millions in fuel bills and provide self-sufficiency both nationally and individually. This is nonsense. If, for instance, only small amounts of excreta are used, the quantity of gas produced will hardly be worthwhile and would probably be less in terms of energy than that required to collect the dung.

The emphasis has now changed, for no longer is methane the sole aim of the process. Other benefits have been shown to be important as well, and there is also a general recognition that problems still remain to be solved. One of the most valuable by-products of the process is the sludge, which retains much of the fertiliser value of raw manure without the latter's unacceptable properties, such as smell. Some people regard the fertiliser value as more important than the methane. Anaerobic digestion is also beginning to be recognised as a more economic form of pollution control than other systems in treating strong agricultural and industrial wastes, for not only is the pollution hazard removed from the waste, but the process is also a net energy producer.

One of the drawbacks is that this type of digestion can not be applied to every case of waste treatment. Each one is different and the suitability of the process has to be measured against the needs of the situation. The system is made to work most efficiently by careful attention to its delicate energy balance, in order to ensure that ultimately there is a net energy profit. The problems which occur are often a result of trying to find the least energy-consuming method of performing such operations as charging and discharging the digester, mixing the waste efficiently and minimising heat losses from the digester. Other inefficiences and difficulties arise with variations in the strength and volume of the waste, indeed one reason why large digesters are considered to be more viable than small ones is that inconsistencies in the waste may be balanced more easily. There are considerable economies of scale, although the process can be made to work in very small units and the smallest viable one may possibly be defined by the smallest use for the gas. However, the smaller the unit, the greater the proportional outlay on accessory equipment, for if these extras are not installed the process becomes more labour intensive per volume of gas produced.

The purpose of this book is to put methane digestion into perspective, and to provide a basis of understanding for both those who are merely interested and those who need their own digesters. It tries to help those who want one to decide what factors should be taken into account in planning and designing their own. Those whose needs are small may be able to build their own with a little extra help and advice from outside, and those whose needs are large will have some grounds on which to judge the ideas and plans put forward by commercial organisations which are trying to sell them a digester.

References and further reading

1. Bell, Boulter, Dunlop and Keiller. Methane — Fuel of the Future. Andrew Singer (1973). Prism Press. (1975).
2. Hobson, Bousfield and Summers. Critical Reviews in Environmental Control. (1974). $\underline{4}$. (2). 131-192.
3. Imhoff, Müller and Thistlethwaite. Disposal of Sewage and other water-borne wastes. (1956). Butterworths.
4. Imhoff. Digester Gas for Automobiles. Sewage works Journal. (1946). $\underline{18}$. 17-25.
5. Parker. The Propulsion of Vehicles by Compressed Methane Gas. Journal of the Institute of Sewage Purification. (1945). $\underline{2}$. 58.
6. Huu-Bang Dao. Production et utilisations du Gaz de fumier. Methane Biologique. CNEEMA Biu° 200. (Sept. 1974).
7. Fry. Methane Digesters for Fuel Gas and Fertilisers. (1973).

2 Understanding the Process and its Requirements

1. Basic Biology

Green plants are able to use the sun's energy directly to form their complex molecules from atmospheric carbon dioxide — their basic cell-building material. In general, bacteria cannot do this, and have to rely on being able to break down these complex plant materials, to extract both the energy bound up in the molecules and the carbon building-blocks necessary for cell growth.

Organic matter, in the natural world, is derived directly from plants, animals or bacteria. It consists of an assortment of molecules whose chief constituent is carbon, associated with other elements such as hydrogen, nitrogen and oxygen. Although the molecules will differ considerably depending upon the form of life from which they are derived, there are three important types of molecules found in all organic matter. These are the proteins which have a characteristic nitrogen content, and whose most important role is as enzymes; the carbohydrates, which are important for maintaining the structure of the cells and as an energy store; and the fats, whose function is primarily energy storage. Both fats and carbohydrates may be considered to be the 'fuel' supplies of bacteria. Most of these natural molecules are made up of repeating units, for example cellulose (a common plant carbohydrate) which consists of many glucose units joined together in a long chain.

$$- [Glucose] - [Glucose] - [Glucose] -$$

When bacteria degrade complex molecules such as cellulose, the first stage consists of splitting the bonds between the units. This is often done by enzymes which the bacterium releases into its environment specifically to do the job, because it is not able to absorb these large molecules directly into itself. Animals, in fact, find it difficult to degrade cellulose, and rely on bacteria living in their guts to release the glucose which they can then absorb.

It is not until the bacteria absorb the units which have been released from the complex molecules, that they can begin to break them down further to obtain the energy contained in their structure and to utilise the fragments to build up complex molecules of their own. This further degradation of almost all units, be they from protein, carbohydrate or fat, eventually leads to the production of acetic acid which holds a key position in the metabolism of the cell. Glucose is broken down to produce acetate (salt from acetic acid) and carbon dioxide (CO_2); fats are degraded by simple release of an acetate molecule from the long chain. There are many different pathways, and they are usually accompanied by a release of energy in a form useable by the cell.[1]

This process of glycolysis (breakdown of glucose) can take place in the absence of air and is common to most forms of bacteria and indeed to yeasts, which ferment glucose to form alcohol and carbon dioxide. In bacteria, however, the acetate which is formed is degraded further to release yet more energy and carbon dioxide. This process requires oxygen (O_2) and it is at this point that aerobic (with oxygen) and anaerobic (without oxygen) bacteria differ. Anaerobic bacteria can make no further use of the acetate formed, since they are not able to use oxygen. Most anaerobic bacteria expel this excess acetic acid into their environment, but there are a small number of species which can utilise the acetate and convert it to methane (CH_4). Although it costs the bacteria a certain amount of energy to produce methane, the methane bacteria rely on acetate or dissolved carbon dioxide as their only source of carbon and as a source of oxidising power instead of oxygen. Oxygen kills all methane-producing anaerobic bacteria. One of the differences between aerobic and anaerobic bacteria lies in the amount of energy which they can extract from their food. The difference can be seen in the diagram *(Fig. 2)*.

Because anaerobes can only obtain a part of the total energy available in glucose, they are less efficient than aerobes, which can utilise it all by breaking down the acetate. As a result anaerobes generally grow and multiply more slowly than aerobes, and the amount of cellular material which they produce is less than aerobes would for the same amount of food. Both aerobes and anaerobes convert dissolved organic matter into the solid material which makes up their cells; it is these cells which create sludge in waste treatment systems. The quantity of sludge from aerobic systems is usually greater than that from anaerobic systems.

When organic matter is digested in the absence of air, the degradation of the large molecules can be followed in three stages. The first is known as liquefaction; the organic matter is usually in solid form and, in order to become available to the bacteria, it has to be broken up by external enzymes produced

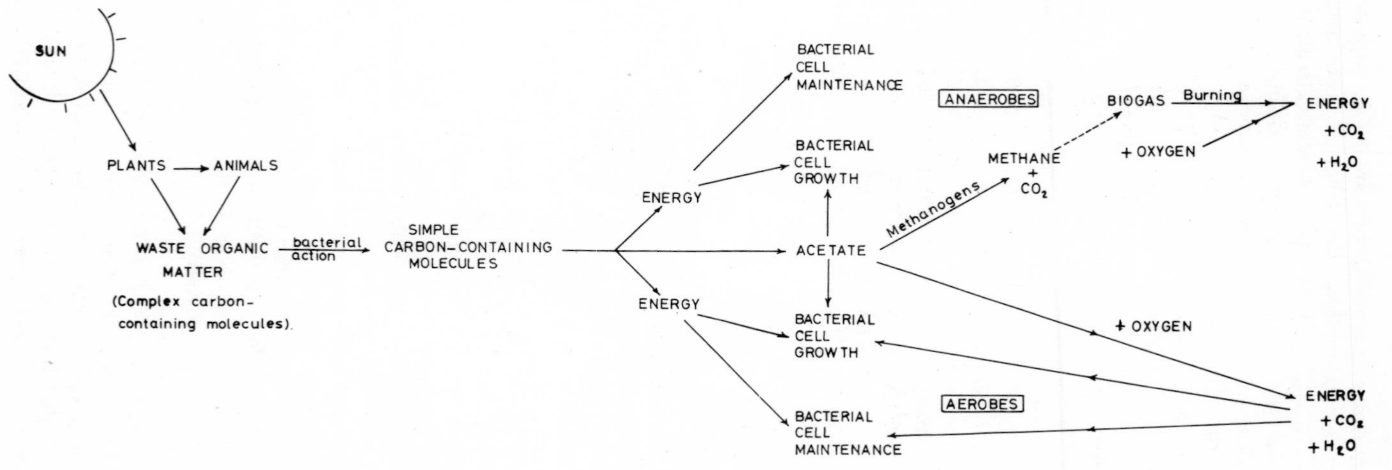

Fig. 2. *Differences in aerobic and anaerobic bacterial breakdown of organic matter.*

8

by the bacteria, and dissolved in the water which surrounds them. It may be difficult to distinguish this stage from what is known as the acid-forming stage, for some molecules will be absorbed without further breakdown and can be degraded internally. The bacteria which carry this out, the acid-formers, can be anaerobic or facultative bacteria (i.e. they can live with or without oxygen). They are important, for not only do they produce the food for the methane producers, but they also remove any traces of dissolved oxygen which may remain in the organic material.

In the acid-forming stage the bacteria produce a few simple compounds which are the end-products of anaerobic metabolism. Apart from acetic, a few other volatile fatty acids such as propionic and butyric are formed. Carbon dioxide and hydrogen (H_2) will also be liberated. At this point most anaerobic bacteria have extracted all that they can from the organic matter and as a result they have to remove their own waste-products from their cells. These substances are shown in the first row of compounds *(Fig. 3)* and their pathways are followed through acetate to methane.

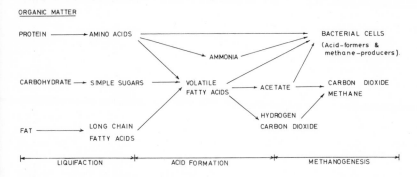

Fig. 3. Breakdown of organic matter by anaerobes.

The methanogenic bacteria take over where acid-formers leave off. There is some doubt as to what end-products of anaerobic metabolism the methane producers can utilise, but it is almost certain that in sewage sludge digestion about 70% of all methane formed comes from acetate and most of the rest from carbon dioxide and hydrogen. At any rate for the practical purposes of running a 'digester' these two sources are the most important. The complete pathway which takes place in a digester can be seen in *Fig. 4.*[5]

The bacterial population involved in the process is very mixed, and this diversity depends to some extent upon the organic material fed into the digester.[3][4] The acid-formers are usually

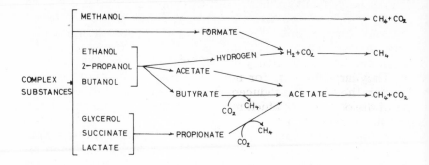

Fig. 4.   Complete methanogenic pathway.

fairly resilient and are better able to withstand sudden changes in conditions than the specialised methane producers. The important fact to note about the bacterial populations in a digester is that they are very interdependent. The acid-formers ensure that the digester contents are free of oxygen. They produce the basic feed material for the methane producers and the action of their enzymes upon proteins and amino acids liberates ammonium salts, which are the only source of nitrogen which methane producers can accept.

In their turn, although they would not be able to survive without the acid-formers, the methane bacteria remove the end-products of acid-formers' metabolism and convert them to gases which escape from the system. If this conversion did not occur in the end, conditions in the digester would become so acid that even the acid-formers would not survive. The result is that in a working digester a balanced population is built up upon an interdependent (symbiotic) relationship between the two bacterial types. If a sudden change in conditions takes place, such as in temperature or in organic content, the first thing which usually happens is a reduction in methane production coupled with an increase in acidity as the volatile fatty acids build up.

2.   What can be digested — nutritional needs of the bacteria

Plant material offers one of the best sources of carbon to both animals and bacteria, because of its high carbohydrate content. It can be digested by bacteria directly, but the process is speeded up if the structure of the plant is broken down first. The most convenient way of doing this is by feeding the plant to an animal first — a process frequently done! The animal's digestive system breaks down the plant both physically and

10

chemically while extracting some of the energy and nutrients at the same time. The animal excretes a considerable amount of organic matter, which it is either unable to use or does not need, being sufficiently nourished. Undoubtedly an anaerobic digestion system would have more organic matter to use if plant material were fed to it, without passing through an animal, but the amount 'lost' to the animal is the price one has to pay to have the feed material suitably macerated.

Since this is normal farm practice anyway, the most convenient raw materials for digestion are often manure from animals such as cows, pigs and poultry. The process of anaerobic digestion has, so far, been most often applied to the treatment of sludge arising from human sewage, probably because sewage is collected on a large scale and physical problems of 'handling' the raw material are minimised. Sewage is over 99% water. The solids which are present settle out at the beginning of a sewage works to form a sludge of about 95% water. It is this sludge which is digested anaerobically. The liquid portion of sewage is then treated aerobically by forcing air to dissolve in the water so as to promote aerobic bacteria to break down the organic matter. The aerobic process produces another sludge. This secondary sludge is separated from the water, and digested with the primary sludge. The water by now has little organic matter left and can be discharged into a river without fear of pollution.

Farmyard wastes, consisting mainly of animal excrement, have a water content ranging from about 95% to 75% and have a higher organic matter concentration than human wastes. The difference between sewage and farm wastes lies mainly in the method of collection — animal wastes are treated as semi-solids, whilst sewage is treated as a liquid. The approximate compositions of two sewage sludges and a piggery waste[1] are shown in *Table 1* for comparison:

Table 1

	Sewage Sludge 1	Sewage Sludge 2	Pig Waste
Carbohydrate	34%	24%	54%
Fats	14%	20%	8%
Protein	19%	21%	21%
Ash (non-organic matter)	35%	28%	18%

Sewage sludges have a higher fat and ash content than animal wastes, whereas the latter have a higher carbohydrate content. This reflects the differences in diet and the fact that sewage sludges contain wastes from cooking and washing, for example, as well as human excreta. It does not really matter about the exact composition in practical digesters, for if the waste comes from a similar constant source a balanced population of bacteria will evolve which is capable of digesting the waste.

The important thing is that there should be enough carbon present. Usually if there is a high concentration of organic solids (about 5%) in the sludge this condition will be satisfied. The solids are also important physically since they provide a surface which these anaerobic bacteria need for support in order to thrive.

Of all the organic matter available in plant material, the carbohydrate, cellulose, is one of the preferred materials for digestion. Of all animals, herbivores alone utilise the cellulose by having evolved a multiple-chambered stomach, the so-called rumen, in which live certain anaerobic bacteria. These bacteria break down the cellulose to acetate which the herbivore can use in competition with some methanogenic bacteria also living in the rumen. Wastes from ruminants do not produce as much methane gas as those from other animals. This is a measure of the success which ruminants enjoy in utilising the bacteria in their stomachs.

Cellulose is normally easily digested by bacteria, but cellulose from some plants is less easily degraded when it is combined with lignin. Lignin is another complex molecule which forms the rigid and woody structure of plants, and bacteria find it almost impossible to break down. If the waste material to be digested contains too much lignin, there may be difficulties with digestion and the woody material will tend to float and form a scum. Straw contains an appreciable amount of lignin and may be a problem.

Apart from carbon, the amount of nitrogen in the waste is most important. All living organisms need nitrogen with which to form proteins. Carbon and nitrogen should be present in the diet in the correct proportions. If there is too little nitrogen, the bacteria will not be able to use all the carbon present and the process will be inefficient in breaking down the organic matter. If there is too much nitrogen not all can be used, so that it accumulates, usually as ammonia (NH_3) and this can kill or inhibit the growth of bacteria, especially the methane producers. It is better to have too much nitrogen in the waste than too little, and even if the C/N ratio is less than the optimum (C/N = 20—30:1) the waste can be digested quite satisfactorily (even down to C/N = 3:1). The ammonium content, however, should be maintained within the toxic limits, but this can be done by dilution with water rather than with additional carbon.

In a similar way phosphate is needed by the bacteria, and although the consequences are not drastic if there is a surfeit, a lack of it will inhibit the process. The optimum carbon to phosphate ratio is about 150 to 1. Sulphur is another nutrient which bacteria require, but since the need for it is less than phosphate most wastes will have enough for their requirements. However, if the waste has too high a sulphur content, sulphate-reducing bacteria will grow and convert sulphur compounds to hydrogen sulphide, a poisonous gas, which can cause corrosion problems when the methane is burnt.[1]

As with other biological systems trace elements such as calcium, magnesium, potassium, zinc and iron are necessary in small concentrations.[1] Cobalt is also known to be needed in some of the methanogenic reactions. As with sulphur these are usually found in sufficient quantities in most wastes. Some industrial wastes may be lacking in certain nutrients and steps would have to be taken to add the required amount, if digestion is to continue efficiently. Vitamins have been added to promote the growth of bacteria in digesters, but no advantage is gained since the bacteria seem to be able to synthesise all the vitamins they need. If some bacteria are not able to do so, there are enough other bacterial types within such a diverse population to provide for their needs. Indeed some experiments have been carried out to produce vitamin B_{12} commercially by anaerobic digestion of industrial wastes.[5]

3. Acidity and pH control

When organic material is allowed to digest by itself as in a batch process (see later) the progress of the digestion can be followed by the changes in the acidity.[6] *(Fig. 5)*. A convenient measure of how acid or alkaline the liquid is, is the pH. Pure water, for example, has a neutral pH of 7. If the pH is below 7, the liquid is acid and if above 7, it is alkaline.

Initially the acid-forming bacteria will be breaking down the organic matter and producing volatile fatty acids. As a result the general acidity of the digesting material will increase and the pH will fall below neutral. After a couple of weeks the methane bacteria will begin to make their presence felt and the pH will begin to rise as the acids are broken down to methane. Another factor which tends to increase the pH during this time is the ammonium content, which increases when protein is degraded. Ammonia (NH_3) dissolved in water is an alkali ($NH_4 OH$). This will tend to neutralise the acid. However, as the pH increases above neutral the ammonia becomes more toxic to the methane producers *(Fig. 6)*.[7]

A third factor which tends to prevent pH conditions in the digesting waste from changing too much (or buffering it) is the alkalinity due to bicarbonate. The bicarbonate ion (HCO_3^-) concentration is directly proportional to the carbon dioxide content in the gas and the pH. Thus if the organic matter is

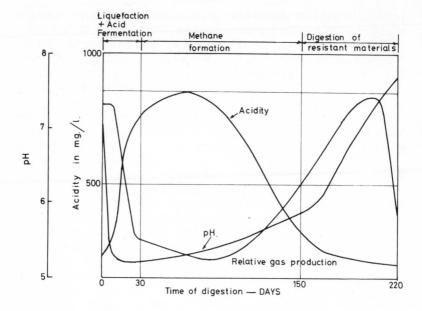

Fig. 5.   Course of digestion of organic solids.[6]

being broken down too fast for the methane bacteria to utilise the acetate and the released carbon dioxide, there will be an excess of carbon dioxide in the gas, and hence a greater concentration of CO_2 dissolved in the liquid as bicarbonate. This will tend to prevent the pH from falling too low which would kill the methane producers, whose optimum pH lies between 7 and 8 (i.e. slightly alkaline).

After a batch digester has been operating for several weeks, the pH rises under these influences and the methane production is at its peak and will balance the organic matter being broken down by the acid-formers. Gradually, however, after several months the organic matter becomes completely degraded and methane production comes to an end. *(Fig. 5).*

The overall picture which emerges is that of closely inter-acting factors which under normal conditions will tend to maintain the pH within the correct range (between 7 and 8) for digestion and methane production. Obviously when starting up a digester the bacterial population will have a greater proportion of acid-formers and a high organic content, which will cause the pH to be low. When the populations are balanced, then the pH will become stabilised at the optimum for the process. However, if the environmental conditions change rapidly, the action of the methane producers would be inhibited and that of

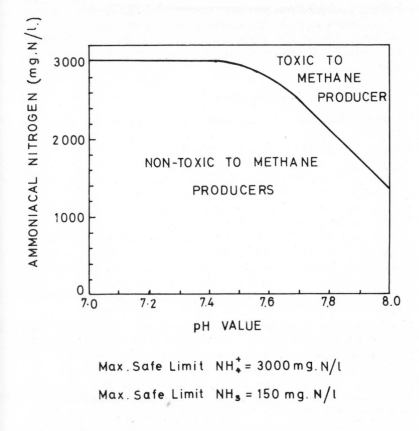

$$Max. Safe Limit \quad NH_4^+ = 3000 \, mg. N/l$$

$$Max. Safe Limit \quad NH_3 = 150 \, mg. N/l$$

Fig. 6. Maximum safe limit for 'ammoniacal' nitrogen.[11]

the acid-formers will be promoted, so that the digester is thrown out of balance.

If the contents of a working digester become too acid, one of the commonest ways of restoring the balance, on a small scale, is to stop feeding the digester for several days.[8] This gives time for the methane bacteria to reassert themselves and reduce the concentration of volatile fatty acids. This procedure is not possible on a large scale such as in a sewage works, where such large volumes of sewage sludge are collected daily that it would be impracticable to store it all. The pH is usually raised by adding lime (calcium hydroxide), which is highly alkaline. This can be disadvantageous if the pH is raised too high, since the carbon dioxide in the digester will combine with the lime to form calcium carbonate, which is very insoluble in water.[7] This not only removes carbon dioxide (a major food source for the

15

methane bacteria) but can also form a scale inside the digester (c.f. in hard water). Scale decreases the efficiency of heat transfer in a heated digester and also takes up valuable space inside which reduces the retention time of the waste in the digester.

4. Temperature and Retention Time

Anaerobic digestion can take place at any temperature between 41-131°F (5°C and 55°C). There appear to be two distinct temperature ranges within this wider range; these correspond to two different sets of bacteria, the mesophiles or those which operate best at 41-104°F (5-40°C), and the thermophiles or those which prefer to live at temperatures between 104-131°F (40°C and 55°C). The rate of gas production increases with increase in temperature, but there is a distinct break in the rise around 104°F (40°C), as this temperature favours neither the mesophiles nor the thermophiles.

In spite of the fact that greater gas production can be expected if a digester is operated in the thermophilic range, this is very rarely done because the energy required to maintain the digester at a suitable temperature more than outweighs the increased gas produced. Also thermophilic bacteria are rather more sensitive to changes in environmental conditions than mesophilic ones, so that the degree of control incorporated in the design would make a thermophilic digester costly. A situation where one might be used, is if the incoming waste is already at a high temperature, such as might be found in an industrial context.

For normal purposes the temperature range considered is 41-104°F (5-40°C). As with most biological processes, the rate of methane production virtually doubles for every 18-27°F (10-15°C) rise in temperature. Not only this, but the total amount of gas produced from a fixed weight of organic waste is also considerably increased, as the temperature of digestion is raised.[9] *(Fig. 7)*. Although it might be expected that a given amount of organic matter would produce a fixed quantity of gas, the increase can be accounted for by a rise in the proportion of acetate being converted to methane rather than being incorporated into the bacterial cell *(Figs. 2, 3)*.

It has been found that the optimum temperature range for digestion is between 86-95°F (30-35°C). This is a temperature range which combines the best conditions for the growth of bacteria and for the production of methane with the shortest retention time of the waste in the digester.

The retention time is a very important factor in digester design. The maximum retention time possible is that which is required to degrade all the organic matter completely. In an unheated digester this might be longer than six months.

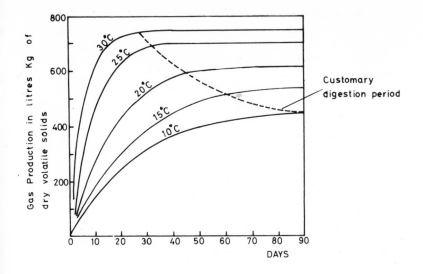

Fig. 7.   Gas production at different temperatures. [9]

However, a compromise has to be found, for as digestion proceeds the organic matter which remains, gets progressively more difficult to digest (the bacteria tend to degrade the easy stuff first).

The retention time of a digester is calculated by dividing the total capacity of the digestion tank by the rate at which organic matter is fed into it. The minimum retention time for the process to work at all is probably around 2-4 days; the reason why shorter retention times are not possible is that the methanogenic bacteria found in digesters have a very slow doubling rate. Some aerobic bacteria have a doubling rate of about half an hour, but anaerobes tend to be slower (probably because they extract less energy from the organic matter they consume). The methane producers double once every 2-4 days, or longer,[7] and if the retention time was shorter than this, the bacteria would be washed out of the tank before they could reproduce, and hence the whole process would stop.

The compromise which is usually reached is to build digesters with a retention time between 20 and 30 days. This has worked well in both sewage works and farm-scale digesters. In sewage works there is often a heated (95°F, 35°C) primary digestion stage lasting about 30 days followed by further cold digestion (normal air temperatures) for about 60 days. At the end of this most of the organic matter has been completely digested.

Digesters have been made to work with as little as five days

17

retention time, but with some farm wastes which are easily biodegradable ten days appears to be optimum,[1] giving good gas production and a reasonable degradation of organic matter. This can only be achieved if the digester contents are maintained at a temperature of about 35°C and if they are completely mixed with the incoming raw waste. If mixing is not efficient, pockets of material at different stages of digestion, different pHs and different temperatures will occur, all of which will impair the rate of the process.

The sensitivity of methane bacteria cannot be stressed enough; their metabolism will be severely affected by rapid changes in temperature. When starting up a digester the temperature should be raised gradually and not more than 2 degrees Centigrade per day.[10] Once the chosen temperature has been reached every effort must be made to maintain it, for variation by more than a degree or so will cause a decline of gas production. Yet again a balance must be found to keep the bacteria in good condition.

5. Toxic materials

Biological systems often exhibit the curious phenomenon of being poisoned by high concentrations of compounds which they need at lower concentrations in order to survive. Methanogenic bacteria demonstrate this by being poisoned by ammonia as was mentioned earlier, particularly when the pH is above 7. It is thought that they are also poisoned by high concentrations of volatile fatty acids (VFA's), their main food supply. This is a matter of some controversy at present, for although some people maintain that it is the VFA's themselves which are toxic above a concentration of about 2000 mg/l (p.p.m.), others say that it is the acidity associated with them which causes the toxity. However, some digesters have not been inhibited by concentrations of acetate as high as 10,000 mg/l (p.p.m.), when the pH is alkaline (i.e. when the acetate is in the form of a soluble salt rather than as acetic acid). This might indicate a reverse sort of relationship between VFA's and pH as exists with ammonium ions. It is difficult to decide which theory is correct, but it is sufficient to say that under conditions of acid pH and high VFA's some inhibition of the process will occur.[1][7][11]

Inorganic salts can complicate the picture also, for strong concentrations of sodium or potassium ions can be inhibiting. It is for this reason that calcium hydroxide is used to raise the pH of a digester, rather than sodium hydroxide, because calcium ions are far less soluble than sodium ones; when they are out of solution, they are no longer toxic.[7][11] *Table 2* shows some of the optimum and inhibitory concentrations of inorganic salts.[7]

Table 2.   Optimum and inhibitory concentrations of inorganic ions in digestion tanks.[7]

Inorganic Ion	Optimum Concentrations	Moderate Inhibition	Strong Inhibition
Sodium mg/l	100-200	3500-5500	8000
Potassium mg/l	200-400	2500-4500	12000
Calcium mg/l	100-200	(2500-4500)	(8000)
Magnesium mg/l	75-150	(1000-15000)	(3000)
Ammonia mg/l	50-1000	15000	8000
Sulphide mg/l	0·1-10	100	200
Chromium % total solids	Not known	2	3
Cobalt mg/l	20	Not known	Not known

Salts of heavy metals, such as copper and zinc, may pose a problem, but under alkaline conditions, and especially if some sulphide is present, these metals will be precipitated and have little effect upon the system. Indeed, although their toxicity may be a matter of a few milligrams per litre under certain conditions, digesters have been run on piggery waste with a copper content of about 80 mg/l and significant inhibition only occurs at about 200 mg/l.[1] If heavy metal poisoning is suspected, the best method of control is to raise the pH by addition of lime followed by the addition of sodium sulphide.[12]

Some organic materials are toxic to methanogenic bacteria. These are usually synthetic materials and include detergents and chlorinated hydrocarbons such as chloroform.[13] [14] Detergents at a concentration of about 15 mg/l have caused difficulties in sewage-works digesters; and in a situation where a sudden spillage or over use of detergent occurs, a small digester may be inhibited because it does not have sufficient sludge to dilute the toxic material, whereas a larger digester would be able to withstand the shock. Stories have been told of how a few litres of chloroform poured down the drain have knocked out the digesters of sewage works treating millions of gallons of sewage each day.

Mention should also be made of antibiotics, disinfectants and pesticides. These compounds can often be found in fairly large quantities on a farm, and care should be taken to prevent them from reaching the digester and killing off the bacteria. This happens all too easily, for instance, after washing down a milking parlour or animal house to prevent infection. It is important to realise that some bacteria are useful, especially those in your anaerobic digester!

The point needs to be made here that small digesters are far more susceptible to sudden changes and to shock loadings of toxic materials. Most digesters, however small, could become acclimatised to a relatively high concentration of toxic material.

If the same concentration were added to an unacclimatised digester it would be severely inhibited. The smaller the digester, the less material is present to "buffer" a sudden increase, whether it be in acidity, heavy metals or disinfectants. Again the need for a regular, consistent feed with constant environmental conditions cannot be over-emphasised.

References and further reading

1. Hobson, Bousfield & Summers. Critical Reviews in Environmental Control. (1974). 4. (2). 131-192. (very good source for further references).
2. Pine. The Methane Fermentations. In "Anaerobic Biological Treatment Processes". Advances in Chemistry Series (1970) 105.
3. Hobson. The Bacteriology of Anerobic Sewage Digestion. Process Biochemistry. (January 1973) 19.
4. Hobson & Shaw. The bacterial population of piggery waste anaerobic digesters. Water Research. (1974). 8. 507.
5. Szemler & Szekelv. Vitamin B12 from Sewage Sludge. Process Biochemistry. (1969). 4. 12.
6. Eckenfelder & O'Connor. Biological Waste Treatment Pergamon Press. (1961).
7. Mosey. Anaerobic Biological Treatment. Institute of Water Polution Control. (1974). Symposium on the treatment of wastes from the food and drink industry.
8. Fry. Practical Building of Methane Power Plants. (1974).
9. Imhoff, et al. Disposal of Sewage and other water borne wastes. Butterworths. (1956).
10. Notes on Water Pollution No 64. Anaerobic Treatment Processes and Methane Production.
11. McKinney. Microbiology for Sanitary Engineers.
12. Mosey, Swanwick, Hughes. Factors affecting the availability of heavy matals to inhibit anaerobic digestion. Journal of the Institute of Water Pollution Control. (1971). 6.
13. Swanwick & Shurben. Effective Chemical treatment for Inhibition of Anaerobic Sewage Digestion due to Anionic Detergents. Journal of the Institute of Water Pollution Control. (1969). 2.
14. Swanwick & Foulkes. Inhibition of Sewage Sludge by Chlorinated Hydrocarbons. Journal of the Institute of Water Pollution Control (1971). 1.

3 Benefits of Anaerobic Digestion

1. Water Pollution Control

Organic matter from human or animal wastes is the commonest pollutant found in water. Organic matter is polluting because the bacteria, which break it down for food, absorb the oxygen in the water. If gross pollution occurs, a stretch of water, be it lake or river, will contain so little oxygen that fish and other aquatic life will be killed and the amenity value of the water will be seriously impaired. The major purpose of sewage works and treatment works is to decompose all the organic matter before the effluent is discharged to the river.

There are two usual methods of measuring the organic content in water: the Biological Oxygen Demand (BOD) and the Chemical Oxygen Demand (COD). The former is a measurement of the oxygen required for bacteria to decompose the biodegradable portion of the organic matter in a fixed period of time (usually five days) and the COD is the oxygen required for all the organic matter to be completely oxidised chemically. The purer the water the lower the BOD and COD. A clean river will have a BOD of between 1-3 mg/l. There are other measurements of water pollution, one of the most important being that of suspended solids (SS). Although many suspended solids will be organic in nature, and hence will have an oxygen depleting effect, large quantities of suspended solids in an effluent can silt up, smother and block channels. The criteria of BOD and SS are often used by Regional Water Authorities in defining the conditions with which effluents discharged to rivers or sewers must comply. Often a discharge to a river must be Royal Commission Standard, namely 20 mg/l BOD and 30 mg/l SS.

Table 3 shows some of the typical characteristics of wastes before they have been treated.

Sewage and industrial wastes are usually treated together where this is possible; some industrial wastes will need pre-treatment before discharge to the sewer because they are so strong that they would overload the municipal treatment works. Agricultural

Table 3.   Characteristics of various untreated effluents.[1][2][3]

Waste	BOD mg/l	COD mg/l	SS mg/l
Sewage	200-400	300-600	100-400
Dairy wastes	800-1500	1500-4000	500-600
Cannery wastes	250-6000	500-15000	20-3500
Meat Packing wastes	700-2000	2000-6000	600-1100
Brewery wastes	400-1500	600-5000	250-650
Farm wastes	500-60000	1000-150,000	800-5000

wastes are rarely discharged to the sewer, and are normally spread on the land without treatment. This can be done without causing water pollution, but the land can only take so much organic matter and at least one acre of land is required for the annual disposal of manure from about 20 pigs,[4] for example. If the land is wet or slopes, the liquid from the manure or slurry may flow into nearby streams or ponds and cause severe pollution; the Regional Water Authorities can prosecute in such instances and demand that steps be taken to prevent such pollution. The Control of Pollution Act 1974 is due to become law by the end of 1975 and far more attention will be paid to the activities of farmers than at present. More pressure may be exerted on those farmers with small amounts of land available for the disposal of increasing amounts of manure, e.g. those with expanding intensive rearing units.

Anaerobic digestion is one of the methods for treatment of strong industrial or agricultural wastes. Organic matter in these wastes is broken down to form methane and carbon dioxide. In comparison with other forms of treatment, such as the activated sludge process, percolating filters and aerobic lagoons, anaerobic treatment has some significant advantages. Aerobic systems, e.g. those found in sewage works, and anaerobic systems both use bacteria to break down the organic matter and in so doing the bacteria grow to produce a sludge which can be easily removed by settlement. Aerobic systems produce far more sludge than anaerobic ones, and so if this has to be disposed of in the end, an anaerobic sludge will cost less to transport.

Anaerobic systems will reduce the organic content of strong wastes (with COD's above 2000 mg/l) more cheaply than aerobic ones; indeed the cost of treating a waste of 20,000 mg/l COD anaerobically has been estimated at about 25% of the cost of aerobic treatment.[5] Anaerobic digestion will deal with wastes with COD's well above 100,000 mg/l. The main reason for this relative cheapness is that there is no need to pump the vast quantities of air into the waste which would be needed for aerobic treatment. Nevertheless many anaerobic treatment processes will reduce the BOD by 70-90% and the COD by 60%.

These reductions may not be quite as effective as aerobic treatment, and the effluents from anaerobic digesters may still be highly polluting (BOD 1000 mg/l). However, since much of the organic matter has in fact been reduced, the pollution hazard of farm sludges disposed of to land will have been minimised. It may be necessary to separate the liquid from the digested sludge, and this supernatant water may need further treatment, which is usually done aerobically, before it is discharged to the river or sewer. Nevertheless the cost of the two systems together would probably be cheaper than a single aerobic system. Undoubtedly if the land conditions can take the sludge as it comes from the digester (i.e. unsettled and non-dewatered) this is the easiest and cheapest method.

Another difference between aerobic and anaerobic treatment is that the fertiliser contents of the raw sludge (nitrogen and phosphate) are retained during anaerobic treatment. Aerobic sludges have a reduced nitrogen content, since much of the nitrogen has been lost with the supernatant liquid discharged to the river. (See later section.)

Then, of course, the greatest advantage of anaerobic treatment is the methane produced. Anaerobic systems are often net energy producers while aerobic ones are net energy consumers. This benefit can be directly offset against capital and running costs of the treatment plant. In summary, a comparison of aerobic and anaerobic treatment systems is shown in *Table 4*.

2. Smell and Odour Control

Smells are one of the most apparent forms of air pollution. Even though they may cause no real damage, people complain more about smells than they do about any other form of pollution. Farmyards have their own characteristic smell and when taken in context, the smell of rotting manure is usually not too objectionable. However, smells cannot be kept in one place and as the size of the manure store increases, so does its smell. The result is that the smells are borne on the wind to nearby human habitation where the complaints arise.

The problem is aggravated by the increasing spread of residential areas and by the increase in intensive animal rearing units, especially of pigs and poultry. The two are incompatible and it is often the farmer who has to give way, who is forced by the local council, under the Public Health Act of 1936, to install some means of controlling the smell, or to close down and take his enterprise elsewhere. The latter course has often been taken, but smell can now be effectively controlled by installing an anaerobic digester.

The smells from decomposing manure are usually caused by the release of such compounds as ammonia, volatile organic

Table 4. A comparison of various aspects of aerobic and anaerobic pollution control systems.

	Aerobic	Anaerobic
1. Character of wastes typically treated	Weak BOD's 100-2000 mg/l Low solids content 0-500 mg/l	Strong BOD's 1000-100,000 mg/l High solids content 2%-20%
2. Reduction in Polluting matter	BOD — 80-95% COD — 70-90%	BOD — 70-85% COD — 60%
3. Quality of Supernatant	Usually low BOD and requiring little further treatment	Higher BOD — may need further treatment in some cases. (Supernatant not necessarily separated)
4. Sludge Production	Larger quantity of sludge produced	Smaller quantity of sludge produced
5. Nutrient Removal	Nitrogen content decreased. Phosphate content the same	Both Nitrogen and phosphate contents much the same
6. Energy use	Net energy consumer	Net energy producer
7. Running cost	Expensive	Less expensive

acids and sulphides. The sulphides, which are probably the most evil smelling, are produced by sulphate reducing bacteria. The smell increases with storage time of the manure, as more of the organic material is broken down. In an open storage tank the smells can escape to the atmosphere and cause complaints. In an anaerobic digestion tank, the smells are contained both in the tank and in the gas holder before the gas is burnt. A well digested sludge has a mild, inoffensive smell, so that even when it is spread on the fields, it causes no complaint. The minimisation of smell is one of the major reasons for digesting sludge anaerobically at sewage works throughout the country.

3. Fertiliser value

Some 99% of all animal wastes are returned to the land in some form, so that some of their fertiliser value is restored to the plants. However, with many disposal methods much of the plant nutrient has been lost before they are recycled, or the nutrients are bound up in a form unuseable by the plants. The potential fertiliser value is enormous, for if all the sewage sludges and all the animal manures in this country were used to the maximum effect, the saving in our £340 million fertiliser bill[6] would be considerable, as shown below:—

Waste from   10 million cattle   @ £6.20 per head   per year = £62 million
Waste from   7 million pigs   @ £1.40 per pig   per year = £10 million
Waste from 120 million poultry @   23p per bird   per year = £28 million
Waste from   60 million humans @ £1.10 per person per year = £66 million

Total value of waste as a fertiliser = £166 million[7]

In many cases it is not possible to use all of these wastes as fertilisers, for the wastes may be too insignificant to warrant collection or may contain toxic materials as do some sewage sludges with a high proportion of industrial wastes. Some piggery wastes contain up to 400 mg/1 of copper which has been used as a feed additive. These toxic materials could harm crops and poison livestock.

The nutrient content of farm wastes will obviously vary depending upon the animals, their diet and the way they are housed. Average values for the percentage compositions of some different wastes are shown in *Table 5*.

Table 5.   *Percentage nutrient contents of various wastes.*[8] [9]

		Nitrogen	P_2O_5 Phosphate	K_2O Potash	
Cattle	Undiluted slurry	0.5%	0.2%	0.5%	
	Farmyard manure	0.5%	0.4%	0.6%	% of wet weight
Pigs	Undiluted slurry	0.6%	0.2%	0.2%	
	Farmyard manure	0.6%	0.6%	0.4%	
Poultry	Undiluted slurry	1.7%	1.4%	0.7%	
	Farmyard manure	1.8%	1.8%	1.2%	
Sewage Sludge		2.05%	4.46%		% of dry weight

Anaerobic digestion of farm wastes does not cause the fertiliser value to decrease. The nutrients present are not degraded in any way, and if the liquid associated with digested sludge is not separated before it is put on the land, the ammonia content is not lost. In all organic wastes much of the nitrogen is bound up in proteins and would not be available to the plants. After digestion at least 50% of the nitrogen present is as dissolved ammonia which is immediately available. Thus digestion increases the availability of nitrogen in organic wastes above its usual range of about 30-60%, depending upon the time of year. The phosphate content is not decreased and its availability of about 50% is not changed during digestion. Potash is usually available at 75 to 100%.[8]

The total and available nutrients in one ton of undiluted slurry are shown below, in *Table 6.*[10]

Table 6. Total and available nutrient contents (in Fertiliser Units) in wastes from cattle, pigs and poultry.[10]

	Total Nutrient Units			Available Nutrient Units			Equivalent fertilizer value (Feb. 75)
	N	P_2O_5	K_2O	N	P_2O_5	K_2O	
Cattle	10	4	10	6	2	10	£1.30
Pigs	13	4	4	6	2	4	£1.15
Poultry	35	30	15	21	15	15	£4.66

The digester of the Rowett Institute treating pig slurry produces a sludge with the following fertiliser constituents:[11]

N.   6.8% = 68 Kg/drytonne or 133 units = £10.64
P_2O_5.   6% = 60 Kg/drytonne or 118 units = £16.50
K_2O. 1.5% = 15 Kg/drytonne or   29 units = £ 1.80
                                                    ‾‾‾‾‾‾‾
                                                    £28.94

The application of particularly strong wastes to the land can result in the 'scorching' of grass, due to the acidity and toxicity of organic matter in such high concentrations. Anaerobic treatment therefore prevents this possible damage by reducing the organic content of the wastes.

However, the organic matter left in the digested sludge can act as a useful soil conditioner, especially for heavy clays and for very light soils. Obviously the type of waste put into the digester will determine how much of a benefit this will be, e.g. straw and bedding material does not digest well and will therefore remain to act as a conditioner. Digested sewage sludge, applied at a rate of 22,000 gallons per acre ($225m^3$/ha) provides the equivalent amount of organic matter as ploughing a one-and-a-half year ley every four years.[9]

In the application of all forms of organic wastes to the land as fertiliser, various factors have to be taken into account as well as the availability of the nutrients to the plants. There are dangers in over-application and nutrient excess. Excess nitrogen makes plants over-vigorous causing 'lodging' (top heaviness) in cereals, and poor quality in potatoes and vegetables. Excess potassium depresses the plant uptake of magnesium, which in turn may cause magnesium deficiency in the animals which eat the plants.

One of the bad effects of excess nutrients is that, if not

absorbed by the soil and plants, they leach into the ground water and are eventually carried into the rivers. This is especially true of nitrates (formed by the oxidation of ammonia) and phosphates. Because they are plant nutrients, they cause the microscopic plants, algae, to grow in the water. This can lead to the problem of eutrophication, when the rivers become clogged with algae. Nitrate is also undesirable in strong concentrations (more than 10 mgN/l) in drinking water supplies, since it causes the disease, methaemoglobinaemia, in young babies.

This problem of leachate is greatest when the ground is saturated with water, for the nutrients will pass quickly into the rivers without being absorbed by the plants. The point that needs to be made is that manure and digested sludge should be applied when they are most effective for the crop requirements. This is a matter for good husbandry and beyond the scope of this book.

In summary, anaerobic digestion does not destroy or lose any of the nutrients from farm wastes, but makes them more available to the plants. It removes organic matter and kills off many pathogenic organisms (see next section); this makes the use of farm wastes less hazardous from a pollution and public health point of view. In certain cases the wastes may need dilution with water (see later) and this may give rise to increased transport costs if the digested sludges have to be moved great distances. It may, however, mean that the wastes are more pumpable and so make spreading easier. As with everything to do with this subject, the circumstances are so varied that every application has to be treated on its own merits.

4. Pathogen Removal

Wastes from both humans and animals contain vast numbers of bacteria, among them pathogenic forms such as the Salmonella types (which cause typhoid in humans and salmonellosis in poultry and cattle) and Brucella which cause brucellosis. Other organisms such as the eggs of Taenia Saginata (the human beef tapeworm) and the potato root eelworm can also be found in these wastes.[12] [13] When they are spread on the land as fertilisers without treatment, there is a serious danger of infection to the animals which graze in those fields. This danger would increase if insufficient time was allowed between application and resumption of grazing. A fallow period of at least one month is said to be necessary to avoid infection. Grazing animals also have their own built-in defence against infection, for they find grass unpalatable where manure or slurry has been recently spread.

However, anaerobic digestion has been found to reduce the

numbers of pathogenic organisms, both bacteria and worm eggs, with the result that application of farm wastes and sewage sludge to the land is less hazardous. The reasons for the reduction in pathogen numbers are complicated, but most significant is the fact that the waste is kept without oxygen for a length of time, between 10 and 40 days, at about 35°C. These conditions may in fact suit the Salmonella organisms, since they are very similar to conditions inside the animals they infect. However, millions of other bacteria in the waste (several hundred million per millilitre) will be competing with the Salmonella and this may be an important factor in reducing their numbers. Normal slurry storage for about a month will reduce about 90% of Salmonellae, but this is due to the fall in pH, as the wastes decompose.[14]

5. Methane Production

The production of methane from organic wastes is the most tangible benefit of anaerobic digestion and in recent times it has been the most emphasised. It is certainly the benefit against which costs are usually measured, if only because the other benefits are almost impossible to quantify.

The biogas which comes from the digester is not pure methane. Its quantity and composition depends upon the feed materials, and may be calculated from Buswell's formula:[15]

$$C_n H_a O_b + (n - \frac{a}{4} - \frac{b}{2}) H_2O = (\frac{n}{2} - \frac{a}{8} + \frac{b}{4}) CO_2 + (\frac{n}{2} + \frac{a}{8} - \frac{b}{4}) CH_4.$$

When this is applied to the three main constituents of organic matter the following quantities and compositions result.

Most organic wastes are made up of a mixture of these three components and the resulting gas composition from their digestion is usually in the range 60-70% methane and 30-40% carbon dioxide. The biogas will contain other gases such as carbon monoxide, hydrogen, nitrogen, oxygen and hydrogen sulphide, the exact proportions of which depend largely on the feed stock to the digester. The presence of nitrogen or oxygen in the digester in abnormal quantities (more than 3% or 0.1% respectively) probably indicates that there is a leak in the system and that air is finding its way into the digester. Hydrogen is a normal product of the acid-forming stage, but since it can be used directly by methane bacteria, its concentration in the gas is seldom very large (between 1-10%). Hydrogen sulphide formation by sulphate-reducing bacteria is governed largely by the concentration of sulphur compounds. Although usually only present in trace quantities, up to 5% can be found in certain wastes (probably industrial rather than agricultural).[16] [17]

Table 7. Theoretical quantities and compositions of gas devised from different classes of organic material.[5]

Material	Composition by weight		Volumes from 1 kg dry material		Volumes from 1 lb. dry material		%
	% CO_2	%CH_4	Biogas	CH_4	Biogas	CH_4	CH_4 by Vol.
Carbohydrate	74	27	0.75m^3	0.37m^3	12.0 ft^3	6.0 ft^3	50%
Fat	52	48	1.44m^3	1.04m^3	23.1 ft^3	16.6 ft^3	72%
Protein	73	27	0.98m^3	0.49m^3	15.7 ft^3	7.9 ft^3	50%

N.B. the difference between the % by weight and % by volume results from the difference in densities of CO_2 and CH_4, being 2 x and ½ x the density of air respectively

The volume of biogas produced per weight of organic matter fed into the digester is also variable. It depends upon many factors, such as the type of digester, the temperature and the rate at which organic waste is loaded into the system. *Table 8* gives some idea of the quantities of gas produced for given farm and industrial wastes and the daily gas production for every animal contributing to the waste.

Table 8.   Quantities of gas produced from some typical organic wastes.[1 2 5 17 18]

Waste Type	Volume of gas/weight of organic material VS			Volume/ animal/day	
	cu.ft/lb.	m^3/Kg	% CH_4	cu.ft	m^3
Sewage sludge	5-12	0.31-0.74	68	1	0.028
Pigs	6-8	0.37-0.50	65-70	8.4	0.24
Cattle	1.5-5	0.094-0.31	65	8.0	0.22
Poultry	5.0-10	0.31-0.62	60	0.5	0.014
Yeast Industry	7.9	0.49			
Meat Packing	8-10.6	0.5-0.66			
Maize Starch	10.7	0.67			
Distillery grain	11.0	0.68			

N.B. The volume of gas produced from cattle wastes is lower than might be expected since a considerable portion of cellulose (the main source of methane) has already been digested by anaerobic bacteria in the rumen of the cow.

The biogas produced can be used directly for heating purposes. In temperate climates about one third of the gas is usually used for heating up the digester contents to its working temperature and maintaining it. Hotter climates obviously have the advantage here and less gas if any needs to be used for this purpose. In general about two thirds of the gas is the net production of fuel from the process. The calorific value of the biogas obviously depends upon the proportions of methane and carbon dioxide, and upon the saturation of the gas with water. The gas contains a lot of water because at digestion temperatures some of the sludge's moisture content will vaporise and mix with the methane. When the biogas contains the normal range of 60-70% CH_4 and 30-40% CO_2, its calorific value varies between 540-700 Btu/ft^3 (20-26 J/cm^3). The net calorific value may be calculated from the percentage of methane as follows[16]:

Net Calorific Value = 8.96 x % methane Btu/ft^3 saturated with water (at 60°F and 30 in Hg)

= (0.33 x % methane Joules/cm^3).

If the gas is dried the calorific value increases, and if the carbon dioxide is removed by 'scrubbing', the calorific value approaches that of pure methane. For comparison of the calorific values of biogas and other gases, *Table 9* shows where methane and biogas stand in the league.

Table 9. Comparison of the calorific values of biogas and other fuel gases.[17]

Gas	Calorific Values	
	BTU/ft^3	Joules/cm^3
Coal Gas	450-500	16.7-18.5
Biogas	540-700	20-26
Methane	896-1069	33.2-39.6
Natural Gas	1050-2200	38.9-81.4
Propane	2200-2600	81.4-96.2
Butane	2900-3400	107.3-125.8

N.B. Variation depends upon degree of saturation and percentage composition of component gases.

Table 10. Volumes of other fuels with calorific value equivalent to 1000 ft^3 of Biogas. (28m^3).[17]

	Fuel	Volume of equivalent fuel	
1000 ft^3 (28 m^3) Biogas @ 600 BTU/cu.ft (22.2 J/cm^3) = 6 Therms (622 Mega Joules)	Natural Gas	571 ft^3	16 m^3
	Liquid Butane	5.3 gallons	24.3 litres
	Petrol	4.3 gallons	19.7 litres
	Diesel Oil	3.8 gallons	17.4 litres

Table 10 shows some of the volumes of other fuels which have a similar calorific value to 1000 cubic feet of Biogas. It is interesting to note that uncompressed biogas occupies a volume which is about 1500 times greater than that occupied by petrol.

These figures show that both biogas and its main constituent, methane, are very viable as fuels. The particular characteristics of methane make it a better fuel for certain uses than others, a factor which will be discussed later. It is important, however, to stress the fact that when mixed with certain proportions of air (5-14% of methane) the mixture is highly inflammable and can easily cause explosions. Extreme care must always be taken to prevent such mixtures from forming, and if they do to avoid their ignition from electrical sparks, matches etc.

6. Self-sufficiency

If we add up the potential biogas produced from all the sources of organic matter available to us in the country, we can estimate the contribution of methane to our national (UK) fuel bill. *Table 11* gives a breakdown of this.

Table 11.   Potential methane production in the UK.

Origin	Numbers	Volume of Biogas		Calorific Value	
		million cu/ft/day	million m^3/day	BTU x 10^6	Mega Joules x 10^6
Sewage Sludge	25 million people (at present treated)	24.7	0.7	14,820	16.1
	(60 million possible)	(59.3)	(1.68)	(35,580)	(38.6)
Pit Wastes	7 million pigs	59.3	1.68	35,580	38.6
Poultry Wastes	120 million poultry	59.3	1.68	35,580	38.6
Cattle	3 million (housed)	23.3	0.66	13,970	15.2
	(10 million possible)	(77.7)	(2.20)	(46,600)	(50.6)
TOTALS PER DAY		166.6	4.72	99,950	108.5
Total per year		60,810	1,723	36,500,000	39,600

N.B.   The figures in brackets are potential only; they are not included in the totals.

The total proportion of biogas could be about 365 million Therms per year in about 61,000 million cubic feet. Our total consumption of gas in this country in 1973/74 was 11,570 million therms; the contribution of biogas to this would therefore only be about 3.2%. This contribution is reduced by between half and a third if some of the biogas is used to heat the digesters, so that only 1.5-2% of our national gas consumption could be supplied by biogas. Natural gas makes up some 14-15% of the total primary fuel required by the country, so that the contribution of biogas to the total fuel requirements would only be about 0.3%.[19]

These figures are the theoretical maxima of gas production,

if all the waste available was digested. In practice the actual amount of waste which can be easily collected for digestion is very much smaller than that shown. Some of the waste is produced in such small quantities, and distributed over such a wide area, that its collection and digestion would cost more than the value of the gas produced.

These figures show that the contribution of biogas to the total energy needs of the country is not large, although any fuel produced here is a saving against energy which has to be imported. However, from the individual's point of view the contribution from the biogas he can create from organic wastes available to him can be considerable, and by substitution of methane for one or two items on his fuel bill he can at least make a significant saving.

It is doubtful whether many individual enterprises could be completely self-sufficient by relying on biogas as their sole source of energy. However, by using the methane most effectively, e.g. for space or water heating, a sizeable portion of the energy bill can be reduced, leaving those functions which have to be powered by electricity to rely on the mains supply.

On the other hand methane can be used to generate electricity. Since the gas is being produced all the time, and a certain amount of it can be stored in gas holders as a reserve, biogas can be used as a reserve energy supply when electricity is cut off by strikes or fuel shortage. This is a very attractive potential for those enterprises (industrial or agricultural) which need a constant regular supply of electricity to run their machinery, e.g. milking and milk cooling. In certain cases this reserve energy could be extremely valuable for emergency uses in hospitals, etc.

It would be extremely wasteful to keep the gas for emergencies only, since facilities for storing all the gas between these crises would occupy unnecessary space. The system which allows the gas to be used for non essential but valuable purposes, such as heating etc. from day to day and for electricity generation in emergency gives both freedom from rising fuel costs and self-sufficiency in the event of power cuts.

References

1. Klein, River Pollution. Vols II and III. Butterworths. (1966).
2. Eckenfelder & O'Connor, Biological Wastes Treatment. Pergamon Press. (1961).
3. Farm Waste Disposal MAFF Technical Report 23. (1970).
4. MAFF leaflet. Livestock Manures — advice on avoiding pollution. (1975).
5. Notes on Water Pollution. (1974). No 64 Anaerobic

Treatment Processes and Methane Production.

6. Cooke. ADAS/ARC Conference on Agriculture and Water Quality, Nottingham. (December 1974).
7. MAFF 'Muck 1975' information.
8. MAFF/ADAS Short-term leaflet 171. (1975). Profitable Utilisation of Livestock Manures.
9. Notes on Water Pollution No 57. Agricultrual Use of Sewage Sludge. (June 1972).
10. MAFF Farm Waste Disposal short-term leaflet 67. (1973).
11. Hobson, personal communication.
12. Hazards in Slurry Disposal. MAFF Leaflet. (1975).
13. Silverman & Guiver. Survival of eggs of Taenia Saginata after mesophilic anaerobic digestion. Journal of the Institute of Sewage Purification. (1960). 3. 345-347.
14. Jones, The effect of storage in slurry on the virulence of Salmonella dublin. Journal of Hygiene. (1975). 74. 65.
15. Buswell and Mueller. Industrial Engineering Chemistry. (1952). 44. 550.
16. Burgess and Wood. The properties and Detection of Sludge Gas. Journal of the Institute of Sewage Purification. (1964). 1. 24.
17. Fry. Methane Digesters for Fuel Gas and Fertilisers. (1973).
18. Baines. Anaerobic treatment of Farm Wastes. MAFF conference on Farm Waste Disposal (Poultry Waste). (1968 (1968).
19. Figures from British Gas Corporation.

4 Practice

1. Different types and adaptations of the Anaerobic Digestion Process

No two uses for anaerobic digestion are the same. Each unit, therefore, has to be adapted to the needs of the situation. There are several basic types of digester to choose from, and each has its advantages and drawbacks. They can be divided into batch types and continuous-flow digesters.

The simplest is the batch digester, in which the organic material is placed in a closed tank and allowed to digest anaerobically over a period of two to six months depending upon the feed material. The contents are usually heated and maintained at the desired temperature; they are agitated occasionally and this releases bubbles of gas from the sludge. The progress of digestion is shown in *(Fig. 5)* p. 14.

This type of digester is very simple to run, since little attention need be paid to it between starting up and emptying out. It is at each end of the process that the real disadvantages occur, for both loading and emptying the waste can be an extremely messy, labour intensive job. Although almost any organic material will digest to a certain extent, the maximum efficiency of digestion can only be obtained if the digester is loaded carefully. Waste space and pockets of air trapped in the sludge should be avoided since these will inhibit the onset of methanogenesis. The carbon to nitrogen ratio also needs to be carefully controlled at the start, since it is difficult to correct once digestion has started.

Because of the waste-handling problems, batch digesters are usually fairly small, although many of the original farm scale units consisted of two or more large batch tanks operated in series.[1] Modern counterparts of this type of digester are still being used both in this country and abroad.

The main use for batch digesters is to assess the digestability of a particular waste before a full-scale unit is built. A typical miniature batch digester has been designed for the Henry

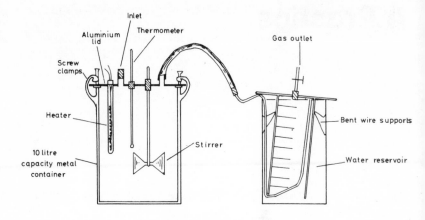

Fig. 8 Experimental batch methane digestor (Henry Doubleday Research Association.)

Doubleday Research Association [4] *(Fig. 8).* This has a digestion tank volume of about ten litres and is suitable for use in schools or laboratories. However, even this size is too large for work in laboratories where many digestion tests are carried out simultaneously. At the Water Research Centre over 20 different wastes are tested at one time using rows of one-litre glass flasks — a quantity which would be impossible using larger volumes. [3]

Biogas Plant Ltd. has produced a larger, dustbin-sized digester (34 litres) *(Fig. 9).* This is similar to, although more sophisticated than, the original L.J. Fry batch digesters made from old fuel drums. [4] [5] Since the 'dustbin' digester may only produce a total of about 50 cubic feet of gas, a single unit may not be very worthwhile. Again the main use for equipment of this size is probably the testing for digestability, although the same information could be obtained from a unit a quarter the size or less.

Batch digesters do have other advantages, for they can be used when the waste is only available at irregular intervals and if it has a very high solid content (25%). If the waste is fibrous or difficult to digest, batch digestion may be more suitable than continuous-flow types, because the digestion time can be increased easily. If digestion does go wrong, due to toxic materials for instance, the batch can be written off and a new one started.

If several batch digesters are used in series, with each at a different stage in the digestion cycle, a continuous flow of gas is obtained *(Fig. 10).* [6] The digesters would be started up at regular intervals, so that as one is approaching the end of its run, a new one is beginning. When this idea is carried to its logical conclusion, and all the batches are digesting in one large tank, this is what is known as the continuous 'plug-flow', or displacement system.

In these continuous-flow types the waste is fed regularly into the digester at one end *(see Fig. 11)*. It displaces from the other end the digested waste which was fed in a month or two earlier. This type of process overcomes the messiness of loading and emptying the system, as in a batch digester. It does, however, require a much more liquid waste feed since the flow through the digester is essential to the process. When loading is carried out regularly, the installation of pumps is more justifiable, especially when dealing with larger quantities of waste. The con-

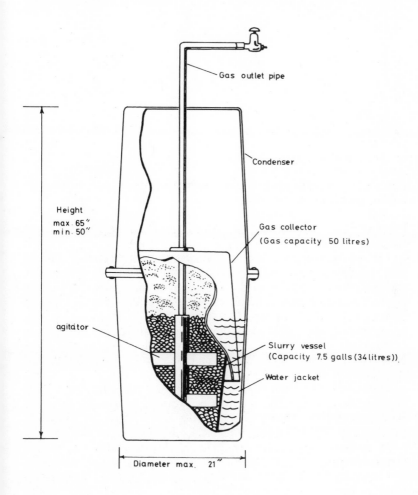

Fig. 9 *Dustbin type batch digester (Biogas Plant Ltd.)*

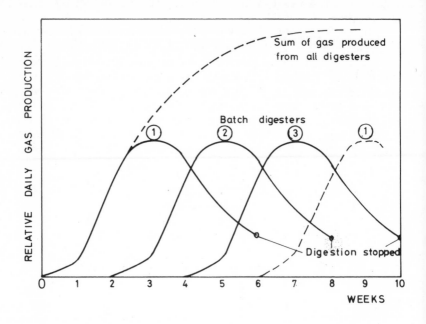

Fig. 10 A continuous flow of biogas from three batch digesters with a staggered operation.

tinuous process is better suited to these larger volumes than batch digesters.

As before, the contents can be heated to increase the rate of gas production, and stirred occasionally. The stirring not only releases the gas by breaking the scum layer which forms (see later) but also mixes some of the anaerobic bacteria into the incoming feed. This reduces the time taken for methane to start being produced by that particular 'plug' of waste. For this reason continuous plug-flow digesters require a shorter retention time than batch units, varying between one and two months, and sometimes less.

There are two main types of plug-flow digester: those that are mounted vertically, such as are found principally in India, developed by the Gobar Gas Institute and its director Ram Bux Singh, *(Fig. 12)* [7] [8] : those which lie horizontally on the ground, such as have been built by L.J. Fry [3] [4] in South Africa and California *(Fig. 13)* and the butyl rubber bag units that are sold by Biogas Plant Ltd. in this country[6] (see Photograph Chapter 9).

Vertical and horizontal designs both have their advantages. The flow of waste into a vertical digester will probably have a

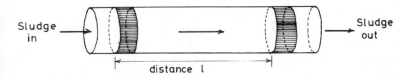

After time, t, plug has travelled distance, l, with no backmixing.

Fig. 11. A diagram to illustrate plug flow in a tubular digester.

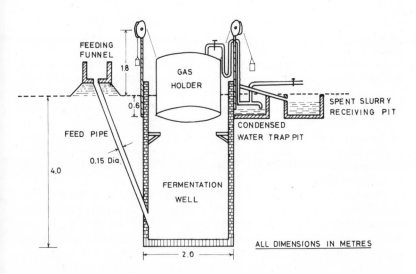

Fig. 12. A diagram of a Gobar Gas Plant (Indian). For 50Kg dung/day. Yield 3 cu. ft. gas/day. Sufficient for the cooking needs of a family 6-8 persons.

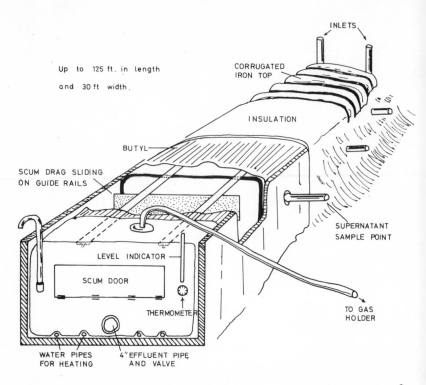

INLETS

CORRUGATED
IRON TOP

Up to 125 ft. in length

and 30 ft width.

INSULATION

BUTYL

SCUM DRAG SLIDING
ON GUIDE RAILS

SUPERNATANT
SAMPLE POINT

LEVEL INDICATOR

SCUM DOOR

THERMOMETER

TO GAS
HOLDER

WATER PIPES
FOR HEATING

4" EFFLUENT PIPE
AND VALVE

Fig. 13. A horizontal displacement digester designed by Fry.[3]

greater mixing effect than in a horizontal one, yet plug-flow will take place more evenly in the latter. The chance of any digested waste passing out before its time is minimised in a horizontal type. However, the retention time in both types is so long (one to two months) and the disruption through mixing so slight that this chance is small even in a vertical digester. A certain amount of mixing is an advantage for 'seeding' the incoming waste with methane bacteria. This is probably the reason why the optimum retention time for vertical digesters is usually shorter than for horizontal ones.

Scum formation is likely to be more acute in vertical units, since they have a smaller surface area than the horizontal type. The scum layer will become thicker at a much faster rate, so that its removal has to be more frequent. Grit removal may be difficult in vertical digesters — especially if they are sunk into the ground — and it may be necessary to empty the tank completely before this can be done. Fry's horizontal digesters have provision for a scum drag and sediment plough to pull both scum and sediment out of one end without disrupting digestion too much.

40

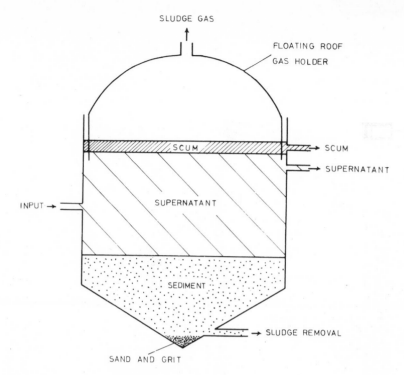

SLUDGE GAS

FLOATING ROOF
GAS HOLDER

SCUM

→ **SCUM**
→ **SUPERNATANT**

SUPERNATANT

INPUT →

SEDIMENT

→ **SLUDGE REMOVAL**

SAND AND GRIT

Fig. 14.   A conventional sewage sludge digester showing the various fractions of digesting sludge.[9]

A vertical digester sunk into the ground will probably require less insulation than a horizontal one, but on the other hand a horizontal unit will not involve the cost of excavating a hole in the ground. The horizontal digester, though, occupies a far larger land area than a vertical one, and this may be an important consideration when land space is limited.

The conventional sewage-works digester is similar in principle to the plug-flow digester and consists of a circular tank in which at least four different functions are supposed to take place. *(Fig. 14)* These are the stabilisation of organic solids by anaerobic digestion (mainly to reduce smell and pollution hazard), the separation of the supernatant liquid, the consequent thickening of the sludge and the collection of methane, which is almost a subsidiary activity. Because they are often unstirred and unheated, conditions in such digesters are less than ideal for anaerobic digestion, which is restricted to the middle third of the tank. The basic design does not allow for very efficient digestion, since the conditions which are best for the individual functions are not necessarily compatible with each other.

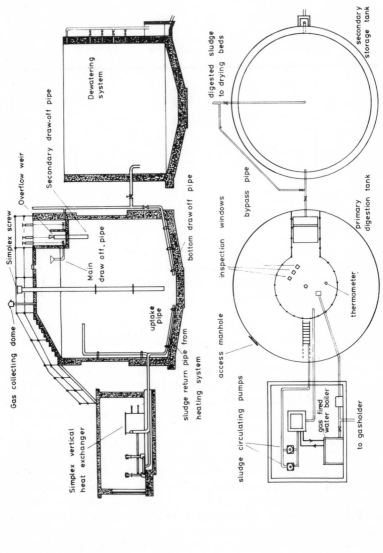

Labels within the figure:

Gas collecting dome

Simplex screw

Overflow weir

Secondary draw-off pipe

Dewatering system

Main draw off pipe

Simplex vertical heat exchanger

uptake pipe

bottom draw off pipe

sludge return pipe from heating system

digested sludge to drying beds

secondary storage tank

bypass pipe

inspection windows

thermometer

primary digestion tank

access manhole

sludge circulating pumps

gas fired water boiler

to gasholder

Fig. 15. Primary and secondary sewage sludge digesters. (Ames Crosta Mills Ltd).

As a result a two-stage process is frequently installed at sewage works, which consists of a high-rate, heated stage followed by a cold, secondary digester.[9] [10] The principal functions of the high-rate process are the stabilisation of the organic solids and the production of methane — these two are compatible.

This first stage requires the incoming waste to be thoroughly mixed with the existing digester contents to prevent thermal stratification — the formation of layers in the sludge at different temperatures — and extensive scum layers. These would tend to slow down the process, which is usually designed to take 10 to 20 days — at least half as long as conventional digesters.

After primary digestion the sludge passes to the secondary digester where stabilisation and concentration occurs under quiescent, unheated conditions *(Fig. 15).* Secondary digesters are often open tanks, since the exclusion of air is not that important at this stage. They can have a retention time of between 20 and 60 days, usually depending upon how the sludge is to be disposed of. Their function as simple storage tanks to ease disposal problems is often reasonably important.

The high-rate digestion system, if suitably modified, can be applied to both agricultural and industrial wastes. If the waste is being produced continually, high-rate digesters are more suitable than the plug-flow type, as displacement digesters are normally fed just once a day or less. For large volumes of waste continual feeding minimises the need for storage tanks. Obviously if the waste is collected only once a day, then either type can be used. In essence, the difference between high-rate and displacement digesters lies in the degree and efficiency of mixing that is required. Efficient mixing minimises scum formation and sedimentation of grit, and in general high-rate digesters produce a sludge with a lower organic content. They are thus more effective as a means of pollution control. Displacement digesters, however, do not need the more sophisticated mixing equipment, so there could be a considerable saving in capital cost.

Much experimental work is going on both commercially and in universities or government laboratories to find the optimum conditions for the high-rate digestion of agricultural and industrial wastes. The most important work emanates from the Rowett Research Institute; after a series of small-scale trials on piggery waste, they have designed a farm-size digester for about 300 pigs. A diagram of this digester shows the layout and equipment required for such a high-rate unit. *(Fig. 16) (see Photograph Chapter 9).*

There are two more recent digester systems [3] [9] [11] known as the anaerobic filter and the anaerobic contact process. These

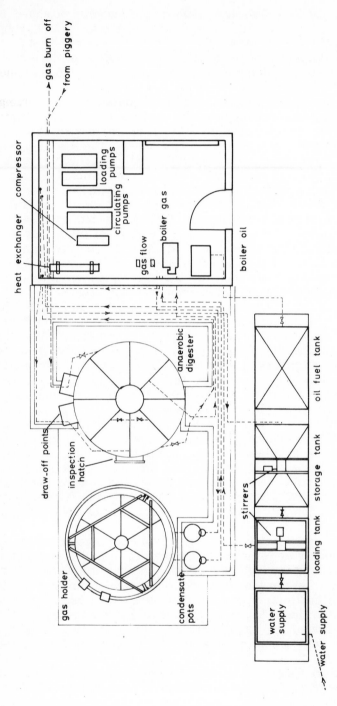

Fig. 16a. Layout of a high rate farm scale digester (Rowett Research Institute).

44

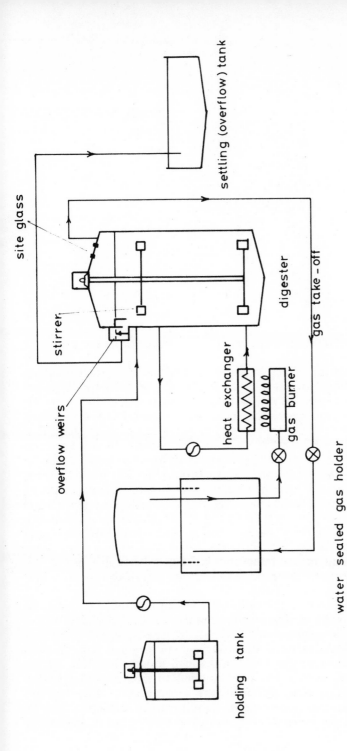

Fig. 16b. Schematic flowsheet for a high rate farm scale digester.

settling (overflow) tank

site glass

stirrer

overflow weirs

digester

gas take-off

heat exchanger

gas burner

water sealed gas holder

holding tank

45

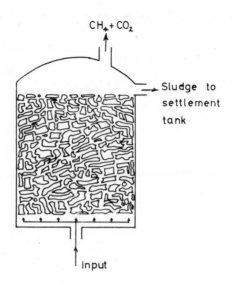

CH$_4$+ CO$_2$

Sludge to
settlement
tank

Input

Fig. 17. Schematic diagram of the anaerobic filter process.[9]

are high-rate in principle and are particularly useful for strong organic wastes with few solids in them. Wastes of this type are usually industrial. Both systems try to prolong the retention time of the bacteria within the digester by retaining or recycling the solids on which they live. The retention time of the liquid is very much shorter (one to two days, or less) than that of the solids and as a result the overall size of the digester can be reduced. In the anaerobic filter system *(Fig. 17)* the digester is filled with small pieces of plastic, upon which the bacteria grow. As the organic matter dissolves in the liquid, they absorb the food they need, break it down and produce methane. Eventually the bacterial growth becomes too large; it breaks off the plastic media and passes out with the liquid as a sludge. This filter system is still in the experimental stage, although work at the Water Research Centre on starch wastes has shown its effectiveness.

The anaerobic contact process relies on the recycling of some of the digested sludge and mixing it with the incoming waste. The bacteria returned with the sludge seed the new feed and digestion starts at once. Because they are recycled, the effective retention time of the bacteria is much longer than that of the liquid. This system has been applied to meat packing wastes with some success, although the separation of the digested solids from the supernatant liquid can be difficult, due to bubbles of gas preventing the efficient settlement of the solids.

Comparison of Different types of Digester system

Type of Digester.	Suitable wastes.	Volumes, Solids content.	Typical Retention times.	Degree of mixing.	Operating Temperatures.	Gas Production.	Degree of control required.	Comments
1. Batch	Agricultrual, Irregular or seasonal. Fibrous or difficult to digest.	Low volumes up to 25% solid.	60 days or more.	Little needed.	Usually 30-35°C.	Irregular and dis-continuous.	Little once started.	Messy and time consuming to start.
2. Plug-flow Horizontal Vertical	Agricultural, continuous or regular flows. Less fibre content.	Larger volumes 5-15% solids	30-60 days	Occasional	30-35°C	Continuous	Simple	Loading and scum removal can be messy
3. Conventional sewage works.	Continuous sewage sludge	Less than 5% solids	30-60 days	Occasional	30-35°C or unheated.	Continuous	Simple	Not very effective
4. High rate sewage digestion								
Primary	Sewage sludge	4-10% solids	10-30 days	Regular	30-35°C	Continuous	More sophisticated.	Automatic
Secondary	From primary Digesters	4-10% solids	20-60 days	None	Unheated	None collected	Simple	
5. High Rate	Agricultural Industrial	4-15% solids	5-20 days	Continuous	30-35°C	Continuous	More sophisticated	Can be automated.
6. Anaerobic contact	Industrial (Agricultural).	Low solids	0.5-5 days*	Continuous	30-35°C	Continuous	Sophisticated	Automatic
7. Anaerobic filter	Industrial	Low solids (low organic contact)	0.5-5 days*	None needed	30-35°C	Continuous	Sophisticated	Automatic

* Liquid retention time.

47

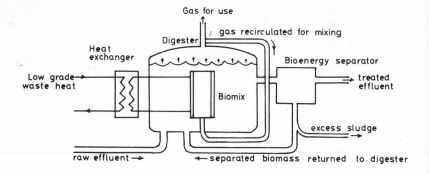

Fig. 18. An anaerobic contact system (Biomechanics Ltd.).

The anaerobic contact process is the basis of the 'Bioenergy' design of Biomechanics Ltd. *(Fig. 18)* which has been successfully applied to starch wastes.

2. Practical points of digester operation

(a) *Starting a digester. Seeding and Sludge Recirculation.* Whichever type of digester is chosen, it needs to be started up carefully. There are two different methods usually employed; the one preferred largely depends upon the type of waste. For batch digesters and for wastes which may not contain the methane bacteria, the usual method employed is known as 'seeding'. This involves mixing the waste with a certain proportion of sludge which has already been digested anaerobically and therefore contains the required bacteria. The higher the proportion of seed sludge, the faster the onset of gas production. According to Fry[4] a useful procedure for seeding is to add a starter of 50% digested sludge and 50% undigested waste. From then on the quantities of undigested waste added each day should not exceed 50% of the total solids content in the digester. This prevents the methane bacteria from being overloaded before they have had time to establish themselves.

The other method, which may also incorporate the addition of seeding sludge[4][12], involves filling the digester approximately 80% full of water and then adding the raw waste regularly (plus seed if necessary) in the volume that it is produced. Provided that the incoming waste is thoroughly mixed with the water, the dissolved oxygen content initially present should be quickly used up and the dilution of the waste should be sufficient to prevent build-up of acidity. This method has the added advantage of replacing most of the oxygen in the digester by the water and there is less risk of explosive mixtures of methane and air being formed. After the first complete retention time has passed, the digester should contain the correct level of solids and the start-up period is over.

When activating digesters for the first time, the most convenient source of seed is from sludge digested at nearby municipal sewage works. After start-up, a digester which is functioning well will not need any further additions of seed, although if it breaks down it may need re-innoculation.

If the waste also needs dilution, the best way of seeding is addition of supernatant; this will contain sufficient numbers of anaerobic bacteria to act as a seed. For high-rate digesters and the anaerobic-contact process, sludge or supernatant may be recycled in order to maintain the concentration of solids in the digester at a constant level. This has the advantage that the system is more precisely controlled and is therefore more stable; however, the equipment required to achieve this reliability may be more sophisticated than necessary.

Whenever starting up a digester, the first portion of the gas produced should always be discarded[4][13], since it will contain some air left behind in the tank, pipes and gas holder. As the gas is discarded it will flush out this residual air and could result in an explosion. Once the gas holder has been filled and emptied several times, we can be sure that no air remains and that the gas can be put to its proper use.

(b) *Loading. Grit removal. Pumps.*
It is very rare that the production of waste occurs so consistently that it can be dosed directly into a digester. In order to balance out the flows a tank is usually necessary; from this the waste is pumped or flows under gravity into the digester. Such a balancing tank is also very useful if, for instance, the digester is overloaded so that feeding has to be stopped or reduced for a day or two. A tank sized to give about one or two day's retention time will allow sufficient flexibility if things go wrong. This size is only practicable, of course, when the daily flows are relatively low.

The balancing tank has other useful functions: it can be used for diluting the waste with water or returned supernatant and, if the waste is too acid, lime can be added and mixed in this same tank. For both of these purposes the balancing tank should be fitted with a stirrer to make sure that the solids are kept in suspension and the waste is well mixed. In certain cases it may be advantageous to use this tank to preheat the waste before it is added to the digester.

Sand and grit can cause problems both in the digester and with pumps, if they are used to move the waste. Grit tends to settle at the bottom of most digesters (to a lesser extent in

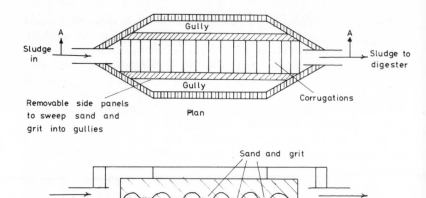

Fig. 19.   Corrugated grit trap (Fry). [4]

high-rate plants) and if allowed to accumulate will take up valuable space. This will reduce the retention time and hence the effectiveness of the system. If possible, grit should be removed from the waste before digestion, and this could be done in the balancing tank. In this case the tank should have a sharply sloping base with a large outlet at the bottom to remove the settled grit. By using a mechanical stirrer in order to move the waste at less than 1 foot/sec, the heavy grit particles will settle while the lighter organic matter will remain in suspension.

There are other more complicated ways of removing sand and grit,[14] but one of the simplest methods is Fry's corrugated-iron grit trap[4] in which the waste flows across the corrugations and deposits the sand etc. in the hollows. *(Fig. 19).* The disadvantage with this is that the surface area is large, and so the waste is liable to oxygenation as it flows over the corrugations, possibly inhibiting methane production.

Once the waste has left the (open) balancing tank, it should not come into contact with air. All pipes carrying the waste into the digester must be air tight, but at the same time easy to dismantle for unblocking and cleaning. The smallest air leak can often severely reduce the efficiency of digestion, as well as causing a potential explosion hazard. Pumps and valves can also be sources of air leaks, and these should be checked regularly as part of routine maintenance. Simple centrifugal type pumps will be quite satisfactory for pumping the waste into the digester, but other types, such as auger or Archimedes screw pumps, can also be used.[14] The latter, although they entrain some air with the waste, are suitable for liquids with a high solids content and are easily unblocked. Both pumps and pipes should have a large bore (a diameter 3″ and over)[15] so that blockages will be minimised. The pumps are rated according to the quantity and frequency of loading the digester. Most pump manufacturers will be able to help with sizing, but ease of maintenance and protection against corrosion are two essentials in any design.

If the waste has a high proportion of large solids, eg. straw, which might cause problems in the pumping or mixing in the digester, these should be filtered out first or broken up in a macerator or shredder. A simple filter of say 1″ mesh over the pump intake will protect the pump and not become blocked too easily; some pumps can incorporate a macerator to cut up large solids and perform two jobs at once. If a waste contains large amounts of straw, a fully-mixed digester will be difficult to operate and a plug-flow or batch digester may be preferable.

Pumps for loading the digester may not be needed at all, and every attempt should be made to use the lie of the land to the best advantage. If you are lucky or hard working, the digester can be gravity fed: lucky, if the slope on the land is sufficient to build a digester below the level of the animal house (either below or above ground); hard working if you are prepared to load up a 'header' tank above the digester. Pumps, however, give a degree of freedom from gravity which can add a great deal of flexibility to the system.

3. The Digestion Tank — Mixing — Scum Control.

The digestion tank itself should, of course, be completely air tight, particularly the inlets and outlets for the waste and gas. This is of prime importance. With open tanks, such as secondary sewage digesters, the scum, forming rapidly, creates a barrier against the inflow of air and the escape of obnoxious smells[14].

The size of the digestion tank is all important. This is determined by the organic loading rate, the retention time chosen and the daily production of the waste[12]. The actual sizing of the tank will be discussed later in the section 'Planning a Digester'. However, other important physical characteristics of the tank should be noted. The digester should have a regular shape, with no nooks and crannies in which digesting material can lodge and not be displaced[9]. This is important both in plug-flow and completely mixed digesters, because if these pockets of material get stuck the overall retention time is reduced. Circular digesters are possibly the best shape and have the largest volume per surface area (a factor which is important for minimising heat loss). Sewage-works digesters almost approach a spherical shape[14], being quite as tall as they are in diameter.

Inside the digester there should be as little equipment as possible in the way of mixers, temperature and pH control probes. If there have to be internal fixtures, these should be designed and sited so as to offer the least resistance to flow and the smallest 'back waters'. Sludge heaters are best placed externally, for the pipes not only create pockets of poor mixing, but also tend to cause the formation of scale, which would impair the heat transfer efficiency.[14]

The continuous digester vessel often has the waste inlets and outlets placed at either end and on opposite sides in order to ensure that the path between them is the maximum. Similar measures to prevent short circuiting are taken with the inlet and outlet for the digesting sludge which is pumped through the external heat exchanger (see later). All inlets and outlets should be so arranged as to preclude any chance of waste siphoning out of the digester and so creating a negative pressure inside. A positive pressure of between 6-8" of water (150-200 mm) is normal. An example of a typical weir outlet is shown in *Fig. 15*, p. 42.

The gas off-take of a digester is usually situated on the top of the roof. With both fixed and floating roofs a pressure-relief valve should also be fitted, with a pipe leading to a burner for excess gas some distance from the digester.

An access hatch is another essential item, for if the digester breaks down, it may need cleaning out through this hatch — a very messy job which should not be undertaken without adequate precautions, e.g. ensuring that digestion has ceased and that the atmosphere inside the digester is safe to breathe and not explosive. Access hatches, as with all points at which the digester wall has been pierced, should be hermetically sealed during normal running.

The two other essential outlets for most conventional displacement-type digesters are the descumming ports and pipes for removal of sediment. High-rate digesters, being continually mixed, are supposed not to have problems of scum and sediment, but usually some does form; designs of high-rate digesters should have points for their removal at both high and low levels. The pipe for removal of sediment can also be used to empty the digester completely, which may be necessary if things go wrong.

Scum is a really big problem in displacement digesters[4][5]. It can take various forms, but usually consists of pieces of hair, feathers, straw, undigested vegetable matter etc. all bound up with grease and fat. Being lighter than the rest of the digesting sludge it tends to float, buoyed up by the gas. Its build-up occurs gradually over several months and it can get so thick that it occupies valuable digester space and reduces the efficiency of organic degradation. When a thick scum layer is broken up, it can release an excessive amount of organic matter back into the digesting sludge and the resulting acidity will impair methanogenesis. Routine agitation to liberate the gases which collects in and below the scum layer will ensure that gas production is reasonably regular. However, the scum layer should preferably be prevented from forming altogether. If this is not done, the layer can build up to such an extent that the gas outlet pipe can become blocked.[16]

The scum should be removed through a fairly large outlet near the top of the digester. Fry[4] has devised a simple scum drag which pulls the scum from one end of his horizontal digester to the 15inch scum port at the other *(see Fig. 13)*. The scum drag is returned to its original position after scum removal for the next occasion on which it will be needed. This type of system has been incorporated into the butyl bag digesters produced by Biogas Plant Ltd. The scum drag, like the sediment scraper or 'plough', is operated on remote-acting lines, which means that gas production does not have to be interrupted while it is being used. Scum removal should be carried out fairly frequently; the timing is obviously a matter of trial and error, for the rate of scum formation will differ from waste to waste.

In sewage sludge digesters a scum duct at the surface is usually incorporated in the design. This may be a simple wide mouthed draw-off point, but sometimes an electrically driven scum board rotates and pushes the floating layer over the edge of the scum duct. Often the scum board is connected to a sediment scraper at the bottom; this pushes the heavy settled solids to a hopper from which they can be removed regularly.[14]

Although provision should be made for scum removal, its prevention is one of the aims of all digester designers. There have been numerous attempts[9] to overcome the problem. Many different types of agitators and mixers have been tried; waste has been sprayed onto the sludge surfaces in an attempt to break up the scum layer; and sludge has been cycled through heat exchangers and gas recirculated in order to obtain more efficient mixing. These ideas have worked at first, but in the end the scum has almost always formed. Of all the possibilities the greatest hope lies in gas recirculation. Even in high-rate digesters, where the contents are supposed to be completely mixed, the problem can still exist. The cycling through heat exchangers does not in itself prevent the accumulation of floating material[15], but together with efficient turbine mixers, or gas recirculation, the scum formation may be considerably reduced or even eliminated.

Efficient mixing is one of the keys to success with high-rate digesters. It was once thought that the presence of extra gas through recirculation promoted methanogenisis, but it now appears that the increase in gas is due to the greater turbulence of this method. Gas recirculation involves pumping some of the biogas back into the bottom of the digester through gas diffusers. *(Fig. 20)* The turbulence caused as the gas bubbles rise to the surface sets up sufficient currents within the sludge to mix efficiently.

Escritt says that a large 75 foot diameter sewage-works digester requires 75 cubic feet of gas per minute to give satisfactory mixing.[14] Obviously smaller digesters will not need as much. The 'Heatamix' system of gas-lift pumps, which circulates the sludge through heat exchangers as well as the digester contents, uses 1.13 cubic metres of gas per minute to lift 6 cubic metres of sludge[17] (40 cubic feet of gas per minute for 212 cubic feet of sludge). The only disadvantage of gas recirculation is that it involves extra handling of the biogas and thus provides an additional potential site for gas leaks, etc.

The other forms of mixing are usually mechanical. A simple paddle stirrer would be neither very efficient nor easy to work in a sludge of say 5-10% solids, but a turbine mixer overcomes these problems and is usually suitable for smaller digesters. Some sewage-works plants circulate the sludge by pumping it through internally mounted tubes. Pumping in a draft-tube is usually carried out by a screw or propeller, which works on the principle of the Archimedes screw *(Fig. 20)*[9] [18]. In general the drive motor of mechanical mixers is mounted outside the digester so that maintenance can be carried out easily.

Such mixers are usually only operated at regular intervals; normally they are used for five minutes every hour, which is sufficient to prevent thermal stratification. On the experimental farm-scale unit at the Rowett Institute this frequency of mixing using two stirrers (top and bottom), each one a third of the digester diameter *(Fig. 16)*, was sufficient to prevent scum formation. The circulation of sludge through a heat exchanger alone could not prevent it.

Mixing or agitation in plug-flow digesters is optional. Fry[4] says that occasional agitation is beneficial, but that regular mixing is detrimental to the process. A horizontal digester is less easy to mix than a vertical one, but in either case agitation by a turbine (hand or mechanically operated depending upon size) is probably sufficient.

4. Sludge Dewatering and Further Treatment

Under normal running conditions, as waste is put into the digester an equal volume of treated material flows out. If loading

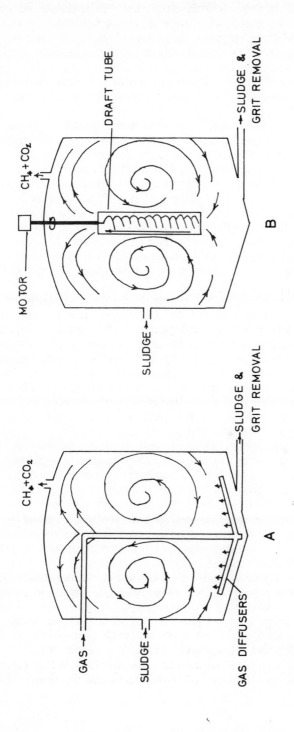

Fig. 20. Schematic diagrams showing mixing by gas recirculation (A) and draft tube mixing (B).[9]

55

is carried out on a daily basis, the effluent can be collected directly into a mobile tanker or muck spreader for immediate disposal to the land. If the digested waste is discharged continuously, or if it cannot be immediately removed, then the effluent will have to be collected into a holding tank.

At sewage works this usually takes the form of a secondary digester, but for smaller units a simple lagoon or balancing tank will be sufficient. The tank should be large enough to cater for the flow accumulated between each disposal operation. If the volume of sludge and its moisture content are too great to be disposed together to the land, the sludge may be concentrated by allowing it to settle in the holding tank. The supernatant liquid can then be drawn off the top of the sludge either continuously or at regular intervals, while the solids remain in the bottom until they can be removed. If the tank is to be used as a means of separating the solids from the supernatant, it may be useful to have a hopper shaped bottom to it, as this will increase the efficiency of separation. The removal of sludge is best carried out through a valved outlet at the bottom of the hopper.

If the sludge does not separate and de-water easily, there are several chemicals which can be added to quicken the settlement. One of the most important of these is lime, which is mixed in surry form with the sludge. Lime is also used as a soil conditioner, so that not only is it readily available to farmers, but it can also be put on the fields mixed with the sludge in one operation. The amount of lime used can be about 5-10% of the total solids[19] . Other chemicals, such as 'copperas' and aluminium chlorohydrate, are sometimes used, but do not have the added beneficial effect.

Reducing the moisture content of sludge in this way lessens the risk of water running off the field and causing river pollution. Also, if a sludge is concentrated it will occupy a smaller volume, and the costs of transporting it will be reduced. One disadvantage of removing the supernatant is that some of the nitrogen is lost with it, so that the fertiliser value of the sludge will be lower. In waste which is particularly strong in nitrogen this could be an advantage, since a more balanced fertiliser (for phosphate and potash) will result.

Another disadvantage of removing the supernatant is that it will probably have to be further treated to remove its polluting contents. Some of the water can, of course, be used to dilute the incoming waste if necessary, but the excess water should either be discharged to a sewer (with permission) or treated aerobically, e.g. in an oxidation ditch — one of the simplest forms of aerobic treatment. This involves extra expense, so that if the land can absorb the sludge as it comes out of the

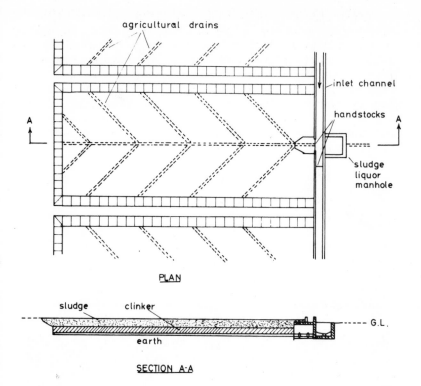

Fig. 21.   Sludge drying beds.[14]

digester, it is obviously better that it should do so.

The moisture content of a concentrated sludge will be in the range of 85-90%. A sludge of this consistency is still slightly runny and can be pumped. If you need a sludge which can be piled in a heap or packed in bags, e.g. for sale as an organic fertiliser, it will have to be dried until it has a moisture content of between 75 and 30%. Mechanical sludge drying, e.g. filter press, vacuum filtration, is expensive both in equipment and running costs, and is probably not worth considering except for large volumes or when a very dry sludge 'cake' is required. The simplest and cheapest method employs sludge drying beds *(Fig. 21)*[14]. Although the most basic form of drying bed is a piece of 'sacrificed' ground onto which the sludge is pumped, a more efficient bed can be made by laying a land drain system covered by a layer of easily permeable material, e.g. clinker, gravel or sand. The sludge is pumped onto this floor to a depth of about 1 foot (250-300 mms) and kept in the bed by low brick walls. Some of the water drains fairly quickly from the sludge (several days) and after about a month of further evaporation the sludge will have a moisture content of 40-50%. At this level it can be stacked up in a stable pile and even bagged for sale as a fertiliser. This is not, however, the most effective method, for the rate of drying is absolutely dependent upon the weather. In hot and dry climates the sludge will dry in less than a month[18], but it could take twice as long or more in wetter climates. However, for most purposes sludge-drying beds will be unnecessary, for the application of a semi-liquid sludge to the land is easier than spreading a solid (which has also lost some of its fertiliser value).

5. Gas Collection and Gas Holders

Unless the gas produced is used immediately or wasted to the atmosphere, it should be stored in some form of gas holder. In this way erratic production/utilisation can be reconciled. The gas holder can either be part of the digester itself, forming a roof floating on top of the waste, or it can be a separate structure connected to the digester by a pipe. This collecting pipe should be situated at least one metre above the level of the sludge or scum[17].

Since the gas, as it is produced, is saturated with water vapour, the collecting pipes and in fact all gas pipes should be fitted with condensate traps. These are placed at all low points in the pipework, for condensate will collect here and if it is not trapped, flow of gas will be blocked. The principle of condensate traps is illustrated in *Fig. 22*. This also shows a way of measuring

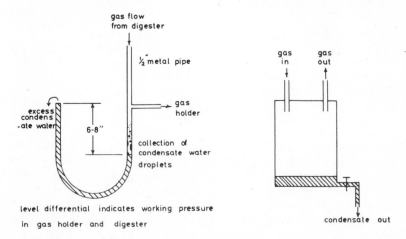

Fig. 22.   Condensate traps (fitted at every low point in the gas pipe system).

the pressure difference between the inside and outside of gas holders. Another important accessory for gas pipework is the non-return valve; this prevents the return flow of gas back up the pipe and also prevents air being sucked into the digester or gas holder via the pipe if an accident happens. *(Fig. 23)*

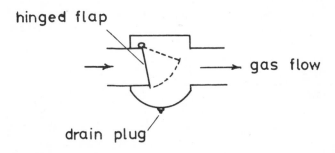

Fig. 23.   Hinged-flap type non-return valve.

It is usual for both floating roofs and separate gas holders to work on the same principle. *(Fig. 24)* When the gas holder is empty, the roof falls under its own weight into the water (or sludge) contained in the lower part of the holder. As the gas is produced, the pressure inside the holder begins to increase. Since the gas pressure is maintained at about 6-8 inches water gauge (150-200 mm), the gas holder rises in order to accommodate the increase in volume. Some gas holders will need counterbalancing weights to prevent too high a gas pressure (see *Fig. 12*). As with the digester the gas holder should be protected against excess pressure with a pressure-relief valve leading to the waste gas burner.

Gas holders are usually circular in shape. The lower water tank can be made of concrete or mild steel; the floating roof is usually made of mild steel, which moves in a rigid frame on rollers. The use of butyl rubber has considerably eased the problems of gas holder construction, for being collapsible it can take the place of a roof floating in water. Two types of butyl gas holder exist. In one the butyl forms the lining of a corrugated steel tank and so works within a rigid structure, and in the other an unsupported butyl pillow bag is filled with gas like a balloon.

As with the separate gas holders, one-way valves on the pipes *(Fig. 23)* to and from the floating holder are important, so as to prevent gas being driven back into the digester or air from being sucked into the holder, should any pressure differences

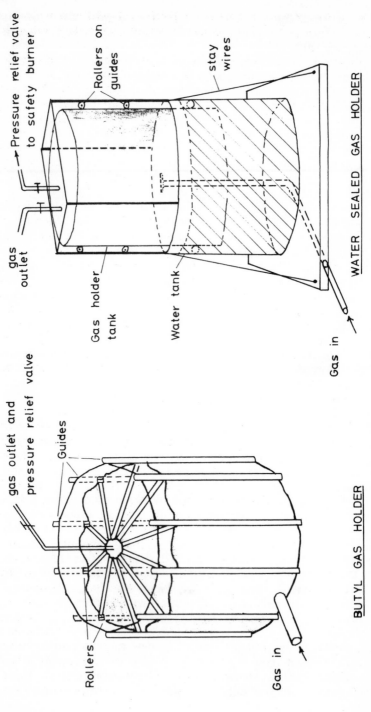

Fig. 24. *Typical arrangements for small gasholders (based on Fry[4]).*

60

occur in the system. Pressure-relief valves should also be incorporated. These allow excess gas to escape when the holder is completely full.

The sizing of the holder depends upon the rate of production of the gas and whether it is to be used continuously or not. If continuously, and the gas use rate is much the same as gas production, only a small holder will be required i.e. to hold the gas output of about four hours. Normal sewage-works practice in the UK is to have holders large enough to store four to eight hours of bio-gas production[17]. If the methane is to be used merely at intervals, e.g. for cooking or as an emergency energy supply, the volume of the gas holder should be larger to take this into account.

Movement of the holder is a very good gauge of the digester's performance. After a digester has been running for some time, its operator will begin to get some idea of how quickly the holder rises under normal conditions. Any fall in gas production — should something go wrong with the digester — will be reflected in a holder that rises more slowly than usual, or even falls, if gas use exceeds output. A gas holder rising faster than normal usually indicates a decrease in the rate of gas use, but more significantly can indicate an air leak into the system, which is giving rise to an explosive mixture with the biogas. Air leaks may be diagnosed by analysis of the gas: a high nitrogen or oxygen content indicates the presence of air in the system.

6. Temperature Control and Insulation

The temperature of most digesters is usually in the range of 30-35°C. It is very important that the temperature, once reached, should be maintained as constant as possible. When starting up a digester, the temperature should be raised gradually at a rate not exceeding 4 F° per day 2° C per day[3]

Heat is needed all the time to raise the temperature of the incoming waste and to replace the heat which is lost from the surfaces of the digester vessel. If we consider the total energy requirements of the system, i.e. the heat required plus the fuel for driving pumps, stirrers etc., and balance these against the gains from the methane, we might easily find that more energy is needed than is produced. For this reason we have to minimise the heat losses and inefficiencies in the whole system. For instance the temperature of animal dung as it is excreted is about 35°C — the optimum for digestion. If the distance and time between the animals' body and the digester can be minimised, the excreta heat loss will be reduced and less heat will be needed to bring the temperature of the waste back to 35°C.

Industrial wastes are often at fairly high temperatures, and the heat can be used to replace losses from the digester surface. Another method of conserving energy is to arrange the inlet and outlet pipes in such a way that heat from the effluent is transferred to the influent. This works on the basic heat-exchanger principle. However, the most important way of conserving heat in the digester is insulation. Commercial insulating materials are usually very adequate for this, but some of the natural materials, such as straw or an earth covering, should be considered for they can be almost as effective and much cheaper.

In some digesters the waste is heated to the correct temperature before it goes into the digester. If the daily volume of waste makes up only a small proportion of the contents, this is probably unnecessary as the cooling effect will be slight.

Heating the influent together with the waste in the digester is the most common way of maintaining the temperature. The simplest way to do this is with a heat exchanger consisting of coils of pipe inside the digester, through which hot water from an adjoining boiler circulates (rather like an indirect domestic cylinder). An example of this is shown in *Fig. 25*, one of Ram Bux Singh's larger designs.

The problem with internal heat exchangers is that they create 'dead' spaces in the digester and scale can form on the pipes which decreases the efficiency of heat transfer. In the end the pipes have to be changed, which involves a complete shut-down of the digester. Scale forms in external heat exchangers as well, but their maintenance is far easier because they are more acces-

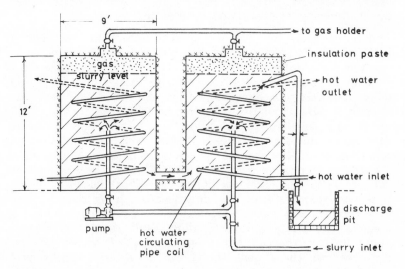

Fig. 25.   A two-stage digester with internal heat-exchanger for 2000-3000 cu. ft. gas/day. (Ram Bux Singh).

sible. In general, heat exchangers for sewage works and high-rate digesters are external (see *Figs. 15, 16 and 18*).

External heat exchangers consist of a coil of pipe with a large bore (4″ diameter) in which sludge circulates. Hot water from the boiler either surrounds these coils or circulates in another coil of pipes, so that the heat is transferred from the water to the sludge. Pumps are needed to circulate the sludge from the digester, through the heat exchanger and back again.

An example of a spiral heat exchanger made by Dorr-Oliver Ltd. for sewage-works digesters illustrates their working principle *(Fig. 26)*. Sludge pumps are not needed for internal exchangers, since hot water will circulate naturally back to the holder once it loses its heat to the sludge. If a faster rate of transfer is needed, water-circulating pumps can be installed.

One method of heating the digester which is neither internal nor involves sludge pumping is that used by Fry and Biogas Plant Ltd. for their horizontal displacement digesters. The pipes carrying hot water from the boiler are laid between the insulation on the ground and the bottom of the digester vessel. Although the efficiency of heat transfer will not be as great as with other heat exchangers, the simplicity and ease of maintenance of this system are important advantages, especially for small digesters.

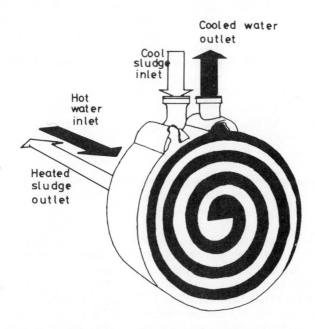

Fig. 26.   A spiral heat-exchanger (Dorr Oliver Ltd.).

The water is usually heated in a boiler, burning some of the gas produced from digestion. This usually amounts to one third of the total methane production. There are, however, other sources of hot water, such as that used for cooling the generator which converts the biogas into electricity. Water can also be heated by capturing the sun's energy with solar panels. Use of hot water from any of these sources liberates the biogas that otherwise would have been used to heat the digester. Since methane is a higher-grade form of energy than hot water, a combination of the two forms of energy-producing systems is a better use of resources. There might, however, be days when solar heating is insufficient and a supplementary source of energy would have to be supplied. There will also be occasions, e.g. on starting up, when no methane is being produced, and the boiler will have to be fired by another fuel, for instance by bottled gas or oil in a different burner.

Probably the form of heating which is most wasteful in terms of energy is electricity using an immersion heater. It is often the most convenient and most suitable form for very small digesters, but the electricity has been generated from a primary fuel such as oil, coal or gas. In this conversion to the secondary fuel, electricity, about 60-70% of the energy has been lost — an unnecessary wastage in these circumstances. These inefficiences are reflected in the high comparative cost of electricity.

Usually heat will be required at a fairly constant rate, but some form of temperature control is advisable. Thermostats placed in the digesting waste can automatically turn the boiler on and off as the temperature falls below or rises above the limits set. If sludge is pumped through an external heat exchanger, one of the best places for the thermostat is at the exit point from the digester. Thermostats or thermometers may also be placed at the end of the heat exchanger or in the body of the digester itself, but for simple manual testing the best place to take the temperature is in the digested waste as it flows out.

7. pH Control

The pH of the digester contents is one of the most useful indications of how the digester is working. When taken with measurements of gas production, any malfunctions in the process can be readily ascertained and action taken.

The pH can be measured electronically with a pH probe mounted inside the digester. This is the most accurate method, for conditions in the sludge change rapidly when a sample comes into contact with air. However, a pH probe and meter are really only feasible in larger plants where the cost of such a system (over £100) would be proportionate to the total outlay.

For small plants measurement of the pH can be adequately carried out upon a sample of the effluent or, if a sampling point has been installed, upon the digester contents. Because the contact with air will tend to alter the amount of CO_2 dissolved in the sludge, the bicarbonate alkalinity will change, the pH will be more acid and allowances should be made for this in the final pH reading. Rapid testing of the sample is therefore necessary. The pH can be determined by using pH papers which change colour when brought into contact with liquids at different levels of acidity or alkalinity. Narrow-range pH papers (eg. from pH 5-9) should be used to detect the pH changes in anaerobic digestion. These can be obtained from a chemist or laboratory-equipment supplier.

The most common pH change is from the normal operating range between 7 and 8, to acid, below 7. As has already been explained, this usually indicates a change in the balance of acid-forming bacteria and methane producers, and is probably best countered by reducing the organic feed for a day or two. If the flow of influent cannot be reduced or if the influent is in itself too acid for good digestion, the addition of lime to the raw waste is often the remedy. The lime must be very well mixed with the slurry before it is put into the digester; the balance tank offers a useful site to add the lime.

It is rare for the pH of the digester to change towards a too alkaline state. When it does, it is probably due to the influent being excessively alkaline. If this is a regular feature, such as in some industrial wastes, the pH should be lowered by the addition of acid. However, if the excess alkalinity in the influent is an isolated occurrence, then the digester should be left to correct itself.

8.   Other controls and analyses

Apart from the total quantity of gas produced, which can be measured by the gas holder or meter, the proportion of methane in the biogas is a useful indication of digester malfunction. Instruments are available (at about £50) which will measure the levels of both carbon dioxide and methane in the biogas. (These or similar instruments will also be useful for testing whether the digester or gas holder is free of gas, so that maintenance or repair work can be carried out both inside and outside the unit.) For small digester units, however, this apparatus will not be justifiable on grounds of cost, and a measure of gas production coupled with a test for flammability (i.e. how it burns) is sufficient.

A measure of the volatile fatty acid content can supplement

pH measurements upon the sludge. A normal digester will operate at a VFA level of 200-400 mg/l, but some wastes may have 4-6000 mg/l VFAs and still be digested without trouble.[9] Methods for the measurement of these can be obtained from standard books on waste-water analysis[20], but this test is more easily carried out in the laboratory.

The total solids (TS) and volatile solids (VS) contents are important measurements, and ones which can be performed at home as well as in the laboratory. The total solids content is that proportion of the waste left after all the water has been driven off. It is measured by weighing a quantity of waste, e.g. 1 Kg, and then drying it out for several hours at about $110°C$ (the temperature of the lowest setting of an oven). The dry solids are then weighed again, and the total solids content and moisture content are calculated as follows:

$$\text{Total Solids} \quad = \quad \frac{\text{weight of dry waste}}{\text{weight of wet waste}} \quad \text{x } 100\%$$

Moisture content = $100 -$ Total Solids %

Volatile solids are those made of organic matter (as opposed to total solids which also contain the inorganic materials, e.g. salts and minerals). In measuring the volatile solids the organic matter is drawn off from the total solids at a temperature of about $600°C$ (in a very hot oven or over a medium gas flame). As before, the waste is weighed prior to placing it in the oven (remember to weigh the empty container first and to subtract its weight). It is left in the oven for about an hour and after allowing it to cool in a dry place (since moisture will be attracted into the waste as it cools), it is weighed again. The volatile solids content is calculated as follows:

$$\text{Volatile Solids} = \text{Total solids\%} - \left[\frac{\text{Weight of dry solids } (600°C)}{\text{Weight of wet waste}} \text{ x } 100 \right]\%$$

The solids remaining after the volatile solids have been driven off at $600°C$ are known as the fixed or inorganic solids.

Digester inputs are usually in the range of 3-10% TS, but an influent with over 10% TS can be used without problems. During digestion the Total solids content is reduced by 40-50%.[9]

The reductions in total and volatile solids contents through digestion can be useful preliminary estimates of the efficiency of the digester at reducing organic matter and hence the pollution hazard. Obviously the more sophisticated tests of COD and BOD on the influent and effluent would be better, but these can only be carried out in a laboratory. As well as the total solids reduction, the appearance of the effluent can be a good indication of digester function. The sludge should be black with a faint tarry, but not unpleasant smell, and the solids should settle out fairly easily.[9] If the sludge smells rancid and foul, something is wrong and further investigations are necessary.

References

1. Huu-Bang Dao, Production et utilisations du Gas de Fumier — Methane Biologique CNEEMA. (1974). Bi No 200.
2. Henry Doubleday Research Assn. pamphlet.
3. Water Research Centre, Notes on Water Pollution No 64.
4. Fry. Practical Building of Methane Power Plants. (1974).
5. Fry. Methane Digesters for Fuel and Fertilisers. (1973).
6. Mitchell. Biogas today — a producers manual. (1975).
7. Gobar Gas Plants. Indian Agricultural Research Institute.
8. Biswas. Fertiliser News. (1974). 19. No 9. 3-7.
9. Hobson. Bousfield and Summers, Crit. Revs. in Environ. Control. (1974). 4. (2). 131 ff.
10 McKinney, Microbiology for Sanitary Engineers. (1962). 247-259.
11. Mosey, Anaerobic Biological Treatment. Inst. Water Pollution Control. (1974).
12. Simpson, J. Proc. Institute of Sewage purification. (1960). 3. 330-336.
13. Institute of Civil Engineers, Safety in Sewers and Sewage Works. (1969).
14. Escritt. Sewerage and Sewerage Disposal. (1965). C. R. Books Ltd., London.
15. Hobson. Rowett Research Institute information sheets.
16. Imhoff et al. Treatment of Sewage and other water borne wastes. Butterworths. (1956).
17. Escritt. Sewers and Sewage Works. (1971). Allen & Unwin.
18. Eckenfelder and O'Connor. Biological Waste Treatment. Pergamon Press.
19. Klein. River Pollution Vol. 3 Butterworth, London. (1966).
20. American Public Health Assn. Standard Methods for the examination of Water and Waste water 12 ed. New York. (1969).

5 Safety

Biogas can cause explosions. The methane content will determine at what proportions, when mixed with air, an explosive mixture will be formed; the lower explosive limit (LEL) for methane is 5.4% and the upper explosive limit (UEL) is 13.9% on a volume basis. Below 5.4% there will not be enough methane present to cause an explosion; above 14% there is too little oxygen to contribute to one. Thus if methane makes up 60% of biogas, the latter's LEL will be 9% and its UEL 23%. The mixture of methane and air which is most liable to explode is that with 10% methane.[1]

The density of biogas, a factor which is important in assessing the dangers of its leakage, varies with its composition. The density of air is 1.29 gms/litre, of methane 0.27 gms/litre and of carbon dioxide 1.98 gms/litre. When the CO_2 content is 45.7% of the biogas the density will be that of air.[1]

The temperature needed to cause an explosion is about 650-750°C.[1] Any spark or lighted match will be sufficiently hot to cause an explosion.

The precautions which need to be taken to minimise the risk of explosion or fire are largely those of common sense. The most probable danger lies in the leakage of biogas. If the leak is into an enclosed space, the probability of explosion is increased. Good ventilation is, therefore, of paramount importance, especially around any parts of the equipment, e.g. sludge pumps, which are housed. If there is good ventilation, the leak of biogas will be quickly diluted by the air, so that the time during which an explosion can take place (i.e. when the mixture is between the explosive limits), is minimised. If the gas is pumped or compressed, for instance in gas recirculation or in the bottling of the gas, the ventilation should be even more efficient. Where large compressors are used, carbon dioxide firefighting equipment should be considered.

Above all no naked flames, lighted cigarettes or shoes which might cause sparks should come into the vicinity of the digester or gas holder. This rule must be rigidly adhered to and a notice warning of the explosion hazard should be posted. It is usually almost impossible to avoid a slight leak of gas, and so the atmosphere around the digester should be treated as explosive as a precaution. Electrical faults and sparks should be avoided, and all electrical installations should comply with the regulations for flame proofing etc. in explosive situations[2]. It is a good idea if pumps etc. can be housed apart from the digester and gas holder.

Areas where the gas is used, for example the domestic boiler system and heat exchanger, should also be separate and housed apart from the digester. It is essential that the pipes leading to the gas boiler (or to whatever use the gas is put) be fitted with a flame trap which prevents possible blow back. Flame traps stop the gas burning further back in the pipe; they consist of a piece of asbestos or metal gauze placed in the gas line. Another simple flame trap can be made by bubbling the gas through a sealed jar of water. Any flames burning in the exit pipe leading from the jar will be quenched by the water[4]. Gas lines should also be fitted with condensate traps, and these and the gas holder should be insulated against freezing up in cold weather.

If a leak does occur on a scale which is liable to cause problems, its presence can be detected by fairly cheap gas alarms which set off a buzzer when the air around them contains more than 25% of the LEL (i.e. 1¼% methane). These instruments are also useful for deciding if an area is safe for the use of welding or electrical equipment in repairing the digester.

When a digester or gas holder needs to be repaired, the gas in it should be purged before any work is carried out. Obviously air must not be used here, for an explosive situation would easily arise. At sewage works the most commonly used purge is the exhaust fumes from a diesel engine, which contain 85.5% nitrogen and 14.5% carbon dioxide[1][2]. Only when the methane has been completely pushed out can air be pumped in to clear out the exhaust fumes. Once this has been done repair work inside the digester can begin. Nevertheless this should still be carried out with care. The safest method of starting or re-starting a digester[2] is to first fill the tank with water to expel most of the air. The daily load of sludge is then inserted until the solids content is up to the operating level. In this way little air remains to mix with the biogas. Nevertheless the first quantities collected in both digester and gas holder must be discharged to the atmosphere, since some air is still bound to be present. Similarly all gas pipes within the system should be purged before attempting to burn any gas for the first time.

Apart from leaks of gas from the digester and gas holder, another danger arises if a negative pressure occurs inside them. Here, air will be sucked in; so in order to ensure that the pressure inside is greater than the atmospheric, piping to and from the digester should not be positioned such that a siphon effect be created, causing a partial vaccuum. Floating roofs have the advantage of sinking onto the digesting sludge, and preventing this from happening. The pressure differences between inside and outside can be varied by the addition or subtraction of weights to counterbalance the weight of the roof. These variations can be easily measured by the levels of water in a U-tube connected to the gas line *(Fig. 22)*. The usual pressure in a gas holder will cause a difference of 6-8" in the levels of water in the U-tube. Both the digester and gas holder should be fitted with pressure-relief valves which should vent to the atmosphere to prevent any build-up of pressure inside them. If the production of gas is greater than its use, the excess gas must be burnt off at some distance from the digester. The excess gas burners may be connected by piping to the pressure-relief valves. These will vent automatically through the burners when the pressure becomes too great. The pressure in the tanks can be lowered manually by by-passing the pressure-relief valves and burning the gas released. Excess gas burners are usually of the Bunsen type and should be large enough to burn the gas at twice the rate that it is produced.[5]

Not only can biogas form explosive mixtures with air, but it can also be asphyxiating when in high concentrations, due to lack of oxygen. If oxygen is less than 17%, the gas becomes difficult to breathe and below 13% it becomes positively suffocating[1]. It is important to discover if there is enough oxygen (O_2) present when any repairs are to be carried out inside a tank.

Some of the other constituents of biogas can be toxic to humans, notably carbon monoxide and hydrogen sulphide (0.1% of CO is fatal in four hours and 0.06% of H_2S is fatal after half an hour). Since biogas can be almost odourless, breathing it unawares can be a danger. However, as far as H_2S is concerned, the 'bad-egg' smell does offer some sort of warning for both this hazard and for leak detection. In order to find actual leaks, the simplest and cheapest method is the use of soapy water which bubbles when spread over the site of a leak. Commercial 'sniffers' can be bought and might be worthwhile for a large plant.

Other points of safety which might be incorporated into the design include the earthing of metal digesters (and obviously all electrical equipment as well). A lightning-conductor could be fitted to a digester if it is large. Water should be available by tap or hose for cleaning, and as a fire precaution. Valves and taps should have some indication of the direction for turning on and off. The whole anaerobic digestion process should be sited fairly well clear of other buildings and, depending upon circumstances, it might be worth considering putting a fence round the site. This will not only deter vandals who might blow up both themselves and your plant(!), but also remind you that you are entering an area where smoking equals stupidity.

References

1. The properties and Detection of Sludge Gas. Burgess & Wood. Journal of the Institute of Sewage Purification. (1964). 1. 24.
2. Institute of Electrical Engineers. B.S.I. No C.P. 1003 — Electrical apparatus and associated equipment for use in explosive atmospheres.
3. Safety in Sewers and at Sewage Works. Inst. of Civil Engineers and Ministry of Housing. (1969).
4. L.J. Fry. Practical Building of Methane Power Plants. (1974).
5. Escritt. Sewers and Sewage Works, metric calculations and formulae. Allen & Unwin, London. (1971).

6 Uses for the Gas

Biogas or methane can be used in just as many ways as town or natural gas. If it is worthwhile installing a digester, it is equally worthwhile finding the most efficient use for the gas. Obviously this depends in the first instance on how much gas is produced. At the same time as considering the uses for the gas, you should take stock of all of your fuel requirements in order to find the least wasteful system of energy use. *Table 12* compares the costs of equivalent amounts of other fuels with biogas to help in calculations.

Methane is a fairly high-grade source of energy; it can provide intense localised heat compared, for example, to energy from solar panels, which provide a generalised warmth distributed by means of water.

One important use which requires high-grade energy is cooking. Each person in the UK needs on average about 5500 BTUs per day (5830 kJ) for cooking purposes. This corresponds to about 9 cu. ft. (255 litres) of biogas and can be produced from the waste of two to three pigs (allowing one third of the gas to be returned for heating the digester). If the gas is burnt in gas ring burners, the energy requirement ranges from 7000 to 10,000 BTUs per hour (7420-10,550 kJ per hour), or, at 600 BTUs/cu ft (22.7 J/cm^3) of biogas, about 11 to 17 cu ft per hour (308-476 litres/hour). Obviously the burners are not used continuously and so gas-holder facilities are necessary to balance out the flows; the maximum hourly consumption of a normal gas cooker is 45,000 BTUs per hour (47,700 kJ per hour) using four ring burners and an oven.[1] According to T.H. Hutchinson of Kenya, his 10-cow units supply 50 cubic feet per day or 30,000 BTUs (1400 litres per day or 31,800 kJ per day).[2] His unit comes with one light and a two-ring cooker. Allowing 25 cubic feet of gas (700 litres) for ten hours lighting, the two burners (7000 BTUs per hour; 7420 kJ per hour) can be used for about two hours per day, which is sufficient for most cooking needs.

Possibly a more efficient cooking use for the biogas is in a

Table 12 *Comparison of Various Fuels with Biogas, showing the energy equivalent price (U.K.) for each (correct to May 1974).*

Fuel	BTU/cu.ft.	BTU/lb.	BTU equivalent of 1000 cu.ft. of Biogas (= 6 Therms)	Price/cu.ft. Price/unit Price/gallon	Price for 6 Therms (= 1000 cu.ft.)
Biogas	600	—	1000	—	—
(Town gas)*	500	—	1200	—	—
Natural Gas	1033	—	581	9.5p/therm	57p
Methane	995	—	603	—	—
Propane	2480	—	242	1.0p/cu.ft. (Bottled) 13.5p/therm (distributed)	£2.42 81p
Butane	3215	—	187	1.25p/cu.ft.	£2.34
Electricity	3411 BTU/unit	—	175 units	1.841p/unit	£3.22
Coal	—	10,200-14,600	59-41 lbs.	1p/lb.	59-41p
Petrol		20,500	29lbs. = 3.9galls.	70p/gallon	£2.70p
Fuel Oil		18,300	33lbs. = 3.5galls.	25p/gallon	87.5p

* Very little Town Gas is now sold, the conversion to Natural Gas being almost complete.

heat-storage type cooker (e.g. an Aga). This is not only a cooker, but also heats water for domestic purposes, or for re-cycling through the digester. This type of cooker uses gas continuously, so that if the rates of gas production and use balance, there may be no need for a holder. However, since there are bound to be occasions when the digester stops or breaks down, provision should be made for piping in an alternative source of energy, i.e. natural or bottled gas. Where such an alternative source is needed, non-return valves should be fitted on both supply lines. An Aga cooker, which also heats 90 gallons (2520 litres) of water per 24 hours to a temperature of about 140°F (160°C), will use 19½ Therms per week or 278,600 BTU/day (292,530 kJ/day).[1] This corresponds to 464 cubic feet of biogas, (13,000 litres) or the methane produced from the waste of 60-100 pigs.

Apart from use in combined cookers and domestic water heaters, biogas can also heat water for central heating. Gas-fired boilers have a similar heat conversion efficiency — of about 75% — as Agas, and the normal range of domestic central heating systems require an energy input of 40,000 to 125,000 BTU's/hour or 67-208 cu ft of Biogas per hour (42,000-131,250 kJ/hour or 1876-5824 litres/hour), i.e. the gas produced from the waste of 250-780 pigs. The use of biogas for central heating may not always be entirely appropriate, for methane will probably be produced all the year round, while a central-heating system will only be needed in the winter. If the biogas is produced from cattle wastes, which can only be collected in the winter when the animals are housed, both the production and the use of the gas would be matched.

Methane can be used to provide direct heating as well, as in a gas fire. Normal energy inputs for this vary from 10,000-20,000 BTUs per hour (10,550-21,100 kJ/hour). A day's methane production from waste of two or three pigs would allow one hour's use of the gas fire. Since this would be an intermittent use of gas, a gas holder would be a definite requirement.

Gas lights usually burn about 2500 BTUs per hour (2650 kJ per hour) for each mantle (4 cu ft per hour, or 112 litres per hour).[3][4] The light produced is not as bright as that from other gases, but the intensity can be increased by passing the biogas over the top of a little petrol, so that some petrol vapour is burnt with the methane. Ram Bux Singh has developed this type of lighting for use in India.

Burners for biogas have to be specifically designed or converted, in the same way that gas appliances had to be converted in the change-over from town gas to natural gas. In order to burn with a more intense heat, air has to be mixed with the gas. The aim of an aerated burner (e.g. Bunsen burner, *Fig. 27)* is to

74

entrain and mix air in the correct proportions with the gas. If the burning velocity exceeds the gas/air mixture velocity, the flame will burn back into the tube. Conversely the flame will lift off the head of the burner if the burning velocity is less than that mixture. The burning velocity depends upon the proportions of gas and air, and the gas/air mixture velocity depends upon the size of the jet and the pressure of the gas.

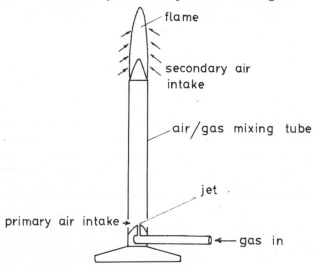

Fig. 27.   Bunsen burner (aerated).

When methane is burnt, it requires 9.57 cubic feet of air per cubic foot of methane in order to burn completely ($9.57 m^3/ m^3$ of methane).[5] In order to achieve this the flame portholes should have a combined cross-section area 300 times the area of the cross-section of the jet. The cross-section of the jet can be determined by how much energy is needed for the particular burner. The air intake orifices are determined by the amount of gas to be burnt. *Table 13* shows how much gas will flow through a square inch or 100 sq.mms. of jet for given gas pressures.

Burners used in any appliances should be protected against corrosion, particularly when raw biogas is used, since it will contain moisture and perhaps some hydrogen sulphide. To remove some of the corrosion hazard, and to increase the calorific value of biogas, it may be 'scrubbed'. This is an essential treatment if the gas is going to be compressed for bottling, for there is no point in bottling a gas a third of which is unburnable carbon dioxide. Scrubbing involves bubbling the biogas through lime water (or a solution of caustic soda) in order to remove most of the Carbon dioxide, followed by

Table 13. *Flows of gas through different jet sizes at different pressures.*[6][7]

Pressure of gas in inches of water	(mms of H_2O)	Cu.ft/min/inch^2 of jet	(m^3/100 mm^2/min)
4	(100)	42½	(0.186)
5	(125)	47½	(0.208)
6	(150)	52	(0.228)
7	(175)	56	(0.246)
8	(200)	60	(0.263)
9	(225)	64	(0.281)

passing the gas over iron filings in order to remove the hydrogen sulphide. When carbon dioxide is absorbed by lime water, the water goes milky; when the lime water is saturated with CO_2, the milkiness clears. In this way you can see when the effectiveness of the lime water has passed its peak and when it needs to be replaced. The amount of lime water needed can be accurately calculated by knowing the rates of flow of gas, its percentage carbon dioxide content, and the calculation from basic principles that 1 gallon of lime water will absorb 0.09 cu.ft. of CO_2 (1 litre of lime water for 0.56 litres of CO_2). If this is from biogas with 30% CO_2 content, the biogas volume would be 0.3 cu.ft. (and 1.9 litres).

The capacity of the iron filings to absorb Hydrogen sulphide is less easy to calculate, since in practice the iron sulphide which appears turns into a sticky sludge which prevents all of the filings from reacting. When this sticky sludge seems to be completely formed, the iron filings need replacing. A larger scale method of scrubbing the gas is to bubble it through water at a high pressure. If you decide to 'scrub', the running costs of buying the chemicals and of their subsequent disposal should be taken into account. For most of the uses of biogas scrubbing is unnecessary.

Although the most direct use of biogas is burning to produce heat, this is not the most versatile form of energy. Biogas may be used to power generators and hence to produce electricity. The efficiency of most generators is only about 25%, but by using their cooling-water to heat the digester, the overall potential may be increased. Generators used are often dual-fuelled and can run on diesel alone or a mixture of diesel and biogas[8]. This is standard sewage-works practice and indeed many plants are entirely self-sufficient in electrical energy.

For generators and stationary engines biogas does not need to be compressed, since the gas can be piped direct to the engine. Normal gas consumption is about 16 cubic feet (450 litres) per horse power per hour[3][4]. However, for automobiles and mobile engines the gas is normally bottled so as to minimise storage space, although Imhoff gives examples of cars and tractors carrying large collapsible bags on the roof.

Methane or biogas does not liquefy very easily (its critical temperature and pressure being -82.5°C (-116.5°F) and 45.8 atmospheres)[9], so that it is not so simple to compress and bottle as some of the other gases, such as propane or butane. In large installations the gas can be compressed to about 350 atmospheres (5000 psi)[10] and stored in special containers. From here it would be transferred to the smaller pressure vessels carried by cars or tractors at about 200 atmospheres (2800 psi) [10]. For smaller installations direct filling of the pressure vessels between 2000-3000 psi will be adequate. A typical container size would be about 5 foot long by 9 inches diameter (1.6 m x 0.27m in diameter), capacity 1.9 cubic feet (54 litres) and weigh about 140 lb (63 Kg). This would hold the equivalent of 420 cubic feet of uncompressed methane, which is comparable to about 3½ gallons of petrol (15.9 litres).[10] Coulthard in Australia uses two 280 lb (127 Kg) gas bottles, with gas at 2000 psi; his car has a range of 100 miles (160 Km)[11]. However, not only does the storage of the gas occupy three to four times the space that petrol does, but also the weight of the cylinders considerably adds to the overall payload of the vehicle. Coulthard has to have a spring fitted under the rear end of the car to compensate for the extra weight of the gas containers. Obviously if you decide to convert your car to run on biogas, the amount of compression and weight of the cylinders will reflect on the alterations to the car and the average length of journey in between filling up. Running a car on biogas becomes a little more worthwhile for short, local trips, when you can fill up with gas frequently (at least daily). Unless the situation changes, your digester will be your only 'filling station' for miles.

There are also inefficiencies in the compression of the gas: it will use up energy (at about 25% efficiency), and as a fuel for the engine the compressed gas will again burn at about 25% efficiency. The actual useful work when gas provides the power for mobile engines must be lower than any other use. Nevertheless, it has been done successfully on a number of occasions — probably the most ingenious and large-scale operations of this sort were the powering of fleets of sewage-works vehicles during the war[10]. The advantage such fleets have over casual users of sludge gas is that the vehicles are confined to one location and pass the fuel store often enough to fill up whenever necessary. Agricultural tractors confined to a farm would be in a similar situation. Yet besides the inefficiencies in this use of the gas, there could be problems arising from its compression, since it involves extra handling and further possible risks of explosion. A Home Office licence has to be obtained in order to store methane in the liquid state (but it is very rarely necessary to compress it that far).

The use chosen for the gas will depend upon the needs of each situation. The domestic uses are obvious and can be applied directly. In the case of autonomous houses biogas represents a supplementary fuel. It is obvious that the wastes from a small number of people, without any livestock, can not produce enough gas to perform any of the principal tasks outlined above — the waste from ten people might produce enough methane to keep a gas ring alight for one hour a day.

In a rural or semi-rural situation the position is different, for not only can the waste from the humans be used (with only a small flush of water to prevent over-dilution), but animal and vegetable wastes could provide the main bulk of organic matter for digestion, so that cooking or water heating can be carried out on a reasonable scale. Some designs which incorporate the digester within the house itself are potentially dangerous and invite the half-serious question of what happens when a lighted match is dropped down the lavatory? In all cases the digester and gas holder should be situated apart from other buildings.

On the farm the gas can be used for many purposes. In intensive animal rearing, heat, either general or localised, is often needed to provide a constant temperature for the growth of young animals. In the dairy industry the gas could be used to power the refrigerators or to generate electricity to power the milking machines. Where grain is produced and stored, the gas can be used to dry and maintain the grain in good condition. Grass, also, can be dried using sludge gas[12].

Similarly horticulturalists can utilise gas to maintain the temperature inside their glass-houses. Many, in fact, use bottled gas already. If direct gas heating is used, the increased carbon dioxide content in the exhaust fumes from the heater would provide a stimulus to plant growth during the day. During the night, however, there might be problems of too high a CO_2 content in the glass-house atmosphere; obviously the system has to be carefully regulated. The carbon dioxide content of the atmosphere, after the gas has burnt, may have to be watched if direct heating is used for both plants and animals: the increased ventilation which may be necessary will raise the fuel consumption, so that the benefit from using methane for direct heating may be reduced.

In industry the major uses for biogas are to supply electricity and/or central heating to offices and workshops. Apart from industries producing large quantities of digestible waste material which can supplement their existing energy sources with their own biogas, there are others who could come to an arrangement with a nearby farmer, or even sewage-works, to provide the waste. The cost of transporting the waste would have to be offset against the benefits in this case; the source of waste would have to be fairly close to the industry. It might even be cheaper to perform the actual digestion on the farm (with the firm's financial backing) and to pipe the gas to the industrial site.

Although the gas can be used in various ways throughout the world, in developing countries the potential qualities of methane become more significant. The energy shortage there is even more acute than in the 'developed' countries, and the smaller the amount of energy available, the more important the contribution of biogas becomes. The annual energy consumption in India is 57.1 Therms per head compared to 1333 Therms per head in the UK ($6x10^9$ Joules per head against $140x10^9$ Joules per head).[13] The success of the Gobar Gas Plants in India bear witness to the need and applicability of this type of village technology.

If large numbers of village or family-sized digesters were built throughout India, the total energy produced from animal and human wastes would be double the amount produced from conventional power stations for the same investment[13] . Not only this, but the improved fertiliser value of the waste would reduce the need for imported and expensive inorganic fertilisers. At present cow dung itself is dried and used as a fuel; cow dung does not have such a high calorific value as methane, and in burning, its fertiliser content is lost. Anaerobic digestion of cow dung, therefore, wins on both counts, and if human wastes are digested as well, the benefit of reducing the risks of disease may be added. Developing countries with fewer sewage works than advanced industrial nations, are in a far better position to make immediate use of human wastes at a local level.

References

1. Commercial Information; Gas & Electricity Boards, Aga.
2. Commercial Information; T.H. Hutchinson, Kenya.
3. Fry. Methane Digesters for fuel, gas and fertilisers. (1973).
4. Biswas. Cow Dung Gas Plant for Energy and Manure. Fertiliser News. (1974). 19. No 9. 3-7.
5. Priestley. Industrial Gas Heating. Ernest Benn Ltd., London. (1973).
6. Escritt. Sewerage and Sewage Disposal. C.R. Books Ltd., (1965).
7. Escritt. Sewers and Sewage Works. Allen & Unwin. (1971).
8. Keep. Some Notes on Dual Fuel Engines and Pumps. Journal for the Institute of Sewage Purification. (1959). 1. 74-81.
9. Burgess & Wood. The Properties and Detection of Sludge Gas. Journal for the Institute of Sewage Purification.

(1964). 1. 24-46.

10. Parker. The Propulsion of Vehicles by Compressed Methane Gas. Journal for the Institute of Sewage Purification. (1945). 2. 58.

11. Water and Waste Treatment. (1975). $\underline{18}$. (3). 12.

12. Leigh. The use of digested sludge and digester gas for the production of dried grass. Journal for the Institute of Sewage Purification. (1949). 28.

13. Pyle. Costs and Benefits of Methane Production in Poor Countries. I.T.D.G. Colloquium on Methane Production by Anaerobic Digestion. (1974).

7 Uses for the Sludge

The digested sludge can be returned to the land in several ways. When it comes out of the digester, the solids content will have been reduced by 40-50%, so that if the input had about 6% solids, the output will have between 3 and 4%. This is a fairly watery slurry which, when compared on the slurry graph *(Fig. 28)*[1], can be seen to be suitable for organic irrigation by pump and manure gun or by tanker. Thus the simplest method of disposal is direct application of the whole waste in the semi-liquid state with no further treatment after digestion.

The advantage of spreading the complete waste on the land (i.e. supernatant and solids together) is that the dissolved ammoniacal nitrogen has not been lost through supernatant removal. This portion of the nitrogen content is still directly available to plants. If just the solids are used, the fertiliser value is reduced.

The disadvantages of using the whole effluent are the possible increased transport costs for a larger volume and, on farms in wet areas, the addition of yet more liquid to the fields, producing a polluting run-off and general boggy conditions. Whole effluent disposal is more suitable for drier farms. The drier the land conditions, the smaller the storage space required for the digested effluent, since the weather will not play such a decisive role in the timing of the disposal.

For wet farms, or for where the slurry has to be transported over long distances, it may be more practicable to separate the supernatant from the solids. Simple sedimentation is probably the easiest way to do this. After the removal of the supernatant the sludge volume is reduced by about 30%[2] and is therefore more concentrated, i.e. sludge with previously 4% solids will now have a concentration of about 6%. A reduction of one third of the volume reduces the day-to-day pumping or transport costs by a similar amount. However, settlement after digestion poses a new problem — disposing of the supernatant, which can still be highly polluting. The actual volume will only be 30% of the whole waste, so that disposal to the main sewer would be

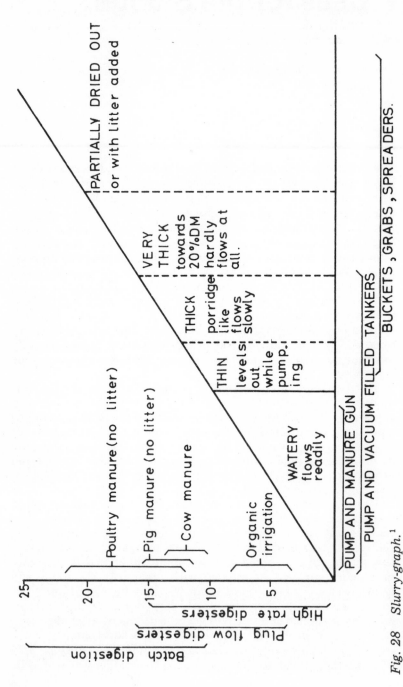

Fig. 28   Slurry-graph.[1]

82

far more acceptable to the Water Authority from both the points of view of hydraulic and organic loading on the treatment works. If a main sewer is not easily accessible, further treatment facilities on the site might have to be installed.

Table 14. Number of animals per land area needed for waste disposal.[2]

Land Use	Dairy Cattle (housed only in winter)		Pigs		Poultry	
	No/Hectare	No/acre	No/Hectare	No/acre	No/Hectare	No/acre
Cut grass	6.2	2.5	100	40	700	280
Grazed grass	3.2	1.3	50	20	550	220
Cereals	1.2	0.5	20	8	220	90
Other arable crops	4.2	1.7	50	20	350	140

The method of sludge disposal to the land obviously depends upon its water content, as can be seen from the slurry graph. Farmers will already have their own methods of waste disposal; if the hydraulic loading on the land is no problem with these existing methods, it is probably best to design the anaerobic digestion system to fit in with them. If the land cannot accept any more water without detriment, the other possibility is to dry the sludge before disposal.

The number of applications of the sludge varies with the land use and the crop to be fertilised. *Table 14*[2] gives an idea of the area needed for the safe disposal of waste from a given number of animals.

These figures are based on undigested waste, but this does not affect the calculations in any way. However, for the best crop results it should be taken into account that more nitrogen will be available in digested waste. A plan of the fertiliser treatments should be drawn up for the different crops. This should include the frequency and quantities of application of the organic fertiliser (from the digester) and the supplementary requirements for inorganic nutrients (from the bag).

The rate of disposal of digested waste probably depends as much upon the water content as the nutrient levels. The land must be able to accept the hydraulic loading without becoming waterlogged or the soil structure 'compacting' under the weight of the vehicles distributing waste on the fields. The rate of irrigation obviously depends upon the crop and the weather. Several inches of water may be needed per acre in one season[3] (there are 23,000 gallons in one acre-inch of water). When slurry is applied by rain-gun waterlogging in many soil types may be prevented by limiting application to 5,000 gallons/acre per week. (50,000 litres/hectare/week). The volume of water is virtually equivalent (90-95%) to the total volume of the waste.

Continual application of undigested sewage sludge and raw animal wastes will often cause an increase in acidity of the soil. This is due to the high organic content in the sludge. For different reasons some artificial nitrogenous fertilisers have a similar effect, and to counteract this acidity lime is often added. The quantities of lime needed may be about four times the weight of nitrogen already applied.[4] Digestion of sewage sludge or animal wastes reduces organic content and hence the tendency to produce an acid soil. Indeed digested sludge may have enough calcium carbonate (lime) alkalinity (12% of dry solids) to satisfy 60% of the lime needed to counteract the acidity from use of normal amounts of nitrogenous fertiliser; the total lime requirement may vary between 0.125 tons/acre and 0.5 tons/acre (0.31-1.26 tonnes/hectare).[4] Thus, although some lime may have to be added, it can easily be mixed with the digested sludge and the two applied together. The addition of lime to the sludge will also condition it and make it easier to settle.

The pH of the soil is important for another reason: the toxicity of heavy metals such as copper and zinc is very much greater if the soil is acid than if alkaline. The concentration of these metal ions in the sludge should be checked periodically, especially if copper and zinc are added to pig and poultry food as growth stimulants. If the sludge is concentrated or dried, the heavy metal content will be correspondingly increased and even more care should be taken. The Ministry of Agriculture, Fisheries and Food has published details of the maximum permissible levels of toxic metals in sewage used on agricultural land.[5]

The sludge can be used for several purposes other than spreading it on the land as a fertiliser. Hydroponic culture of plants involves their growth in a nutrient rich solution rather than in soil; this can achieve a greater productivity than conventional plant culture. Digested sludge, with much of its nutrients in solution, offers an ideal medium for hydroponics. The increased plant growth can provide animal feeds or more raw material for the digester.

This leads on to another use, for which digested sludge is held in lagoons. Algae grow on the nutrients in the solution and when sufficient quantities have been produced, they are gathered, concentrated and fed back into the digester to produce more methane. This is the basis of the sludge → algae → methane system championed chiefly by Golueke and Oswald in California. The emphasis in this system is undboutedly upon methane production rather than treatment of farm wastes. The climate in California is rather more suitable for the high rates of algal production important for this type of system.

There is a second system involving the feeding of fish on the algae produced from sludge lagoons. As before, these algal systems work better in sunnier climates, and tropical fish, for instance, Tilapia, are better suited to rearing in water with a higher organic content than edible fish in this country eg. trout. Carp could be cultured in this way, but the market for this fish in the UK is very limited, although this is not the case in Eastern Europe where carp are a great delicacy. Fry[6][7] goes in to these systems in greater detail and provides references for further reading.

References

1. MAFF (ADAS). Slurry handling, useful facts and figures.
2. ADAS. Advisory leaflet. Livestock manures, advice on avoiding pollution.
3. MAFF. Thinking Irrigation. Farm water Supply leaflet No. 2.
4. (June 1972). Agricultural use of Sewage Sludge. Notes on Water Pollution 57.
5. MAFF/ADAS Permissible levels of toxic metals in sewage used on agricultural land. Advisory paper No. 10.
6. Fry. Methane Digesters, for fuel, gas and fertilisers. (1973).
7. Fry. Practical Building of methane power plants. (1974).

8 Planning a Digester

1. Objectives

In planning a digester we must have very clear objectives. This will enable us to choose the right type of digester and the criteria on which it must be designed. In the first instance the decision must be made as to whether anaerobic digestion is being considered primarily as a pollution-control method — with the production of a usable gas and a sludge with a valuable fertiliser content as by-products — or whether there is greater emphasis on the methane production. Although optimisation of both are not incompatible goals, if methane production alone is important the design criteria are likely to be less exacting and the digester system cheaper to install. If pollution control is the main consideration, the cost of a digester can be offset against the cost of other treatment equipment which might have been necessary. In this case the efficiency of the process may be dictated by the requirements of the local authority or Water Authority; in the case of methane production being the more important benefit, the potential use of the gas will provide the criteria on which to base the design. If more gas is needed than can be produced with existing sources of waste, either the latter will have to be increased, eg, by rearing more animals, or the energy requirements provided in part by other fuels.

Plug-flow, displacement digesters take a longer time to break down the organic matter (measured by the volatile solids or BOD contents) than the completely mixed, high-rate digesters. This can be seen in *Figure 29* in which the reduction in V.S. content is plotted against time for both kinds of digesters. It can be seen that high-rate digesters do not, however, necessarily produce more gas per weight of volatile solids, but because of the shorter retention time the gas is produced more quickly. The total daily gas production is probably very similar for most digester types treating the same waste, since high-rate digesters contain a smaller amount of volatile solids, while systems with longer retention times have a larger total weight of volatile solids but a much lower rate of gas production per day.

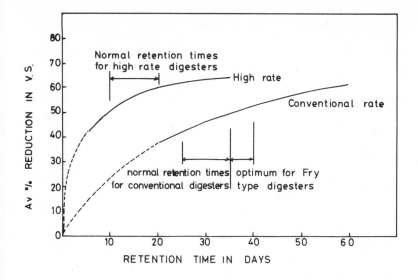

Fig. 29.   Average % reductions in volatile solids with retention times.[12]

2.   (a) How much waste is produced?

Before going any further we must estimate how much digestable waste is going to be available each day. Some ranges of daily volumes of excreta produced per animal or person are given in *Table 15*; the actual figures will depend upon a number of factors, including the age and size of the animals and what they are fed on. The total volume of waste will depend on the number of animals being reared, how they are kept and how much water is used for washing down their living area.

Thus in estimating the volume we must take into account whether the animals are housed day and night all year round, or whether, as in the case of most dairy cattle, they are housed only in winter. In summer the only cattle waste which could be gathered easily is that dropped in the collecting yard and milking parlour. If animals are only housed at night there will probably be half the daily quantity or less.

Secondly the calculations should take into account the way in which the animals are kept. If they are housed on slats, the waste will be almost entirely made up of animal excrement plus a small amount of spilt food. This waste is liable to be more watery than if the animals are kept on straw or other bedding materials, which will tend to absorb the moisture and bind the dung. Undoubtedly slurry without bedding is easier to digest (especially in completely mixed systems), although

87

Table 15. *Volumes or weights of excreta produced per day by different animals*[1-6]

Animal (type of food)	Body Weight lbs.	Body Weight Kg.	Volume of Excreta/day Gallons	Volume of Excreta/day Litres	% faeces	% urine	Ratio of total daily waste/Body Weight
Cattle							
Dairy cow (silage and concentrates)	1000+	454+	7-10	32-45	70	30	7.2%
Fattener (silage and barley)	1000	454	6	27			6 %
Beef fattener	450	203	2½	11			6 %
Pigs							
(Dry meal fed)	100	45.4	1	4.5	45	53	10 %
(Pipeline fed)	100	45.4	1.5-2	6.8-9			9 %
(Whey fed)	100	45.4	2-3	9-13.5			12 %
	40-80	18-36	0.6 (5.6lb)	2.7			10 %
	80-120	36-54	1.2 (11.5lb)	5.4			
	120-160	54-72	15 (14.6lb)	6.8			
	160-200	72-90	1.8 (17.6lb)	8.2			
Poultry							
Layers	5	2.3	0.25lb	0.11 kg	—	—	5 %
Broilers	3	1.4	0.1lb	0.05 kg	—	—	3.3%
Geese/Turkeys	15	6.8	0.5lb	0.22 kg	—	—	3.3%
Horses	850	383	5	23	70	30	6 %
Sheep	67	30	0.5	2.3	66	34	7.5%
Humans	150	68	0.3	1.4	20	80	2 %

the added bedding material may increase the carbon content and hence total gas production. Since bedding materials often contain a lot of lignin, which is almost impossible to digest, the problems involved from scum formation may not be worth the extra gas that is produced. On top of this there may be difficulties in transporting or pumping a waste with a lot of bedding in it.

The incidental water which finds its way into the waste may constitute a significant proportion of the total. It comes mainly from washing down animal houses, dairies and milking parlours to clear the dung or to clean up the floor after the excreta has been removed mechanically. With cattle waste an allowance of between 1-5 gallons/head/day (5-23 litres/head/day) should be made for both yard and milking parlour washing, so that the total volume of waste produced per cow may vary from about 12-20 gallons/day (54.5-91 litres/day). Similarly with pigs the water usage may be about 2 gallons/pig/day (9 litres/pig/day); the total volume of waste then being 3-5 gallons/pig/day. (14-23 (litres/pig/day).[7]

The other sources of diluting water come from spillages from the water troughs, from taps being left running and from rain water. A normal tap running at full bore will deliver about 6 gallons of water per minute (27 litres/minute) or 7200 gallons per day ($32.7m^3$ /day), or enough to dilute the waste from 700 cows by half. One inch of rain water falling on 1000 square feet of yard or roof produces 500 gallons (one centimetre of rain on 100 square metres gives one cubic metre). Unless we need to dilute the waste, it is best to keep all rain water collecting systems separate from organic-waste systems. If dilution is necessary, the correct amount of rainwater can then be carefully added. Other sources of water for dilution are, of course, the mains water (costing 50p to £1.00 per 1000 gallons in some rural areas[8]), and (less pure) water from general yard or dairy washings.

Thus the first stage in our calculations provides us with knowledge of how much waste is produced, at what times of day and whether it is produced all the year round. We also know what diluting factors need to be considered and whether the total waste contains excreta plus bedding material or not. The figures given in the table are only guide-lines and since every farming situation is different, they must be verified or corrected as far as possible before a practical digester is built.

If human wastes are being considered as well, it should be remembered that with the present form of sewage removal (ie. being carried by water) about 40 gallons of water are used per person per day. The resulting strength of the waste is thus far lower than animal wastes, even though a large quantity of

Table 16. Quantities and concentrations of some characteristics of excreta produced daily by different animals.[1-6,10]

Type of waste	Daily Volume Gals.	Lit.	Moisture Content%	Total solids lb.	Kg.	Conc.	Volatile solids %TS	lb.	Kg.	Conc.	B.O.D. lb.	Kg.	Conc.	Total Nitrogen %wet weight lb.	Kg.	Conc.	C/N Ratio	Population Equivalent
Humans	0.3	1.4	89	13	.13		84	.25	.11		.18	.08	.17	.17	.008		6:1	1
Diluted and plus. cooking wastes	40	182	99.9			750 ppm				630 ppm			450 ppm			50-80 ppm	3-6:1	
Cattle.	8	36	87	10.4	4-7		80-90	8.8	4		1.3	.6		0.5	0.04	0.02	15-20	5-8
Diluted x 2	16	72	93.5			6.5%				5.2%			8100 ppm			1100 ppm		
Pigs.	1	45	89	1.6	.7		83	1.3	.6		.44	.2		0.5	0.05	0.02	6:1	2-2.5
Diluted x 3	3	14	95			5%				4.4%			14,700 ppm			2000 ppm		
Poultry.	.03																	
1000 birds	30	136	75	63	28		76	48	22		17	7.7		4	1.8		5:1	100
Diluted x 4	1206	545	94%			6.2%				4.8%			17,000 ppm			4000 ppm		

wasted food material is carried in the sewage as well. If water is to be conserved, a much smaller flush for lavatories could be used, and this would also result in the waste being stronger and more suitable for direct anaerobic digestion. Bath and washing-up water etc. could be plumbed into a separate disposal system, so that the digester's contents would not be further diluted. Waste food and kitchen refuse could be ground up or added directly to the system for digestion with the rest of the organic matter.

The daily volumes of Industrial wastes can be calculated from knowing how much water is used to make one unit of production or to process a given weight of material. Total water usage is known far more accurately than on farms, for the water authorities usually install water meters in factories so that they can make accurate charges for the water used. However, there are often discrepancies in water use in many factories, and a survey of the wastes is necessary to verify the figures and to pinpoint the sources of water wastage.

(b)   How strong is the waste?

The characteristics of the waste should then be estimated and verified by further analysis. *Table 16* will help in the preliminary estimates of the amount of each characteristic produced per animal per day, while an example of typically diluted excreta gives the concentrations to be expected in actual slurries or manures.

The actual concentrations will vary considerably with age, size and food of the animals as do the daily volumes of waste produced. This will have an effect upon the daily gas production — an example which illustrates this is the reduction in the average calorific value of sewage sludge gas by 30 BTUs/cubic foot during World War II in the UK. This fall can be attributed to a reduction in food wastage and in the lower calorific values of foods during this time[9].

The moisture content, total solids and volatile solids are the easiest to measure and the most important for calculating the size of digester needed. The moisture content is needed since it shows how fluid the mixture is *(see Fig. 28 slurry graph)* and how much dilution is necessary for the best results in anaerobic digestion (90-98% moisture). Having chosen a working moisture content, we can calculate how much dilution water to add, if necessary, and thus the total volume of the waste. Fry[6] states that the best moisture content, is when the slurry has the consistency of cream.

The total solids plus the moisture content make up the whole waste. The volatile solids are expressed as a percentage of the total solids, and from this the total weight of volatile solids can be calculated. The total daily volume of the waste and the total daily weight of volatile solids provide the figures on which the design of the digester is based.

The BOD produced per animal and its concentration in the waste are useful parameters, for they are also an indication of the organic load upon the digester.

The amount of BOD produced per animal is sometimes used to calculate the Population Equivalent (PE) of each animal — a figure which is useful in order to obtain an overall idea of the amount of polluting material in the waste. For instance a herd of 100 cows is the equivalent of 730 people. Sewage-works designers sometimes use population equivalents in sizing various parts of the works, including anaerobic digesters, but since digesters for treating farm wastes have slightly different criteria, PE is not all that useful for this purpose.

The BOD figures also give a measure of the total biodegradable carbon in the waste. A rough estimate of the total carbon content is obtained by dividing the BOD by 1.7. This is useful in the determination of the carbon/nitrogen ratio. This should be in the region of 20-30 to 1 for optimum digestion. The C/N ratio should never be above 30:1, but digestion will take place satisfactorily if it is below this figure. Most animal manures have a low C/N ratio, especially if urine — which contains much of the nitrogen — is collected with the faeces. However, animal wastes usually have quite a high proportion of vegetable matter, either from bedding material or foodstuffs, mixed in with the excreta. Since these have a lower nitrogen content the result is that many slurries or farmyard manures have a reasonably high C/N ratio. The nitrogen content and C/N ratios of some plant matter are given in *Table 17*.

Table 17. Nitrogen contents of some typical Plant materials. [5 6 11]

Plant material	Total N % Dry weight	C/N Ratio
Straws	0.5-1.1	50-150
Hay	2-4	12-20
Grass clippings	4	12
Non-legume vegetables	2.5-4	11-19
Sawdust	0.1	200-500

Some industrial wastes, especially those from the food industry, may contain very little nitrogen and phosphate, eg. starch wastes from potato processing and from starch-reduced food manufacture, malting wastes from breweries and distilleries, and vegetable-processing wastes. Their very high carbon content will need to be balanced by some nitrogen, and it might be possible to do this by adding some of the high nitrogenous wastes from abattoirs and meat-processing factories. Blood and meat wastes often have C/N ratios between 3 and 5 to 1.

Table 18. BOD's and Nitrogen contents of some organic materials.

	BOD's	Nitrogen	C/N
Whey	60,000	0.11%	30-40:1
Silage liquor	40,000	0.22%	11:1
Blood	120,000	10-14%	3-4:1

Apart from the C/N ratio the nitrogen content of wastes is also important from the point of view of ammonia toxicity. The toxic limit for ammonia is about 3000 mg/l (provided the pH is above 7.0); the nitrogen content of the waste should be kept well below this figure. The most practicable way of doing this is by dilution. Thus the waste may have to be diluted on two counts — one, the solids content may be too high and secondly, the ammonia or total nitrogen contents may be toxic. A balance should be found so that the criteria of correct moisture and low enough nitrogen content are satisfied by the dilution.

Thus the second stage in our plan for a digester has been to decide how much dilution of the waste will be necessary, whether to include any additives to correct the C/N ratio and finally, to fix an estimate of the total daily volatile solids content.

3. Sizing the digester

There are three ways in which digesters can be sized: by having a fixed retention time and knowing the total daily volume of influent; by having a fixed loading of volatile solids per volume of the digester; and by having a fixed volume for every population equivalent in the waste.

Usually a combination of the first two methods is used. Different types of digesters have their own optimum retention times to achieve sufficient breakdown of the volatile matter and the maximum gas production. It would seem that in general the variation in the loading rate between the different types is not that great. An analysis of the design criteria of some working digesters is given in *Table 19*, to show how they differ. The retention time varies between about 10 and 40 days depending upon the system, but the organic loading only varies between 0.05 and 0.2 lb/VS/cu ft/day (0.8-3.2 KG VS/m^3/day) for most purposes and is generally around 0.17 lbs VS/cu ft/day (2.8 kg

Table 19. Operating parameters of various digester systems.[5 6 9 10 12-17]

Type of Digester System	Operating Temperature °F	Operating Temperature °C	Retention Time Days	Loading Volatile solids/ Volume/day lb/cu.ft/day	Loading Volatile solids/ Volume/day Kg/m³/day	Operating Moisture content %	Gas production cu.ft/lbVS	Gas production m³/KgVS	Rate of Gas Production cu.ft/lbVS/day	Rate of Gas Production m³/KgVS/day	Gas Production = /retention time
1. Conventional sewage Digesters (Displacement type)	Ambient Cold Digestion		55-75	0.082	1.33	97-94	—	—	—	—	—
2. Standard sewage works digester.	Heated 95	35	25-30 11-46 (average 27)	0.16-0.21 0.045-0.2	2.6-3.4 0.73-3.3	97-94	7-9	0.43-0.56	0.28-0.36		0.017-0.022
High rate Optimum sewage	95	35	15 12-15	0.32-0.71 0.15	5.2-11.5 2.4						
3. Rowett Research Piggery Wastes (High rate)	95	35	10 7 5	0.16 -0.23	2.6- 3.7	95-94	6.8 6.13 6.3	0.42 0.38 0.39	0.68 0.68 0.63		0.042 0.054 0.039
	86	30	10				6.3	0.39	0.27		0.017
4. Auchincruive	Heated		23	0.2	3.2	90	6.3	0.39	0.27		0.017
5. Imhoff mixed farm wastes	Heated		36	0.17	2.8	86-90	5	0.31	0.13		0.008
6. Gobar Gas Plant (Ram Bux Singh)	77- 95	25- 35	28	0.18	2.9	91-93	3.8-8.8	0.24- 0.54	0.14-0.31		0.009- 0.02
7. Fry Original Expts. Piggery	95	35	c.60	0.18	2.9	86-88	7.5	0.47	0.13		0.008
Suggested optimum			35-40	0.17	2.8				0.19		0.012
8. Biomechanics High rate, slude - Recycle, Piggery	Heated		5-10			84-90	8	0.5	0.8-1.6		0.05- 0.1
9. Meat Packing	91	33	2	0.08*	1.3		7.5*	0.47	3.8*		0.24
Yeast Manufacture			2	0.11	1.8		4	0.25	2		0.12
Maise Starch	73	23	3.3	0.11	1.8		7.1	0.44	2.2		0.14

*Loading rates are measured in lbs. BOD/cu.ft/day
Gas Production: cu.ft/lb BOD.

VS/m^3/day). This, then, is more important a criterion than retention time, for retention time can be easily altered by adjustment of the moisture content.

The third method is probably not very useful for purposes other than the design of sewage-works digesters and even then it is open to errors. As a point of interest, cold digesters have been designed on 6-12 cu ft of digester volume per person and heated digesters on 2-5 cu ft per person.[4]

Having evaluated the waste as it arrives for digestion, for daily volume, moisture content, total and volatile solids produced per day, if we take a figure of about 0.17 lbs VS/cu ft/day (2.8 kg/m^3) as being around the correct loading for a digester, we can calculate the minimum volume necessary to digest that weight of volatile solids. In choosing a figure of 0.17 lbs VS/cu ft/day (2.8 kg/m^3) as a preliminary design loading we can be sure of sizing the digester in approximately the correct range. It is better to oversize than to undersize it, for this not only allows for an increase in the amount of waste produced but also for estimating errors, and ensures that the waste will be sufficiently digested. It may be an advantage to find a range by calculating the sizes with loadings of both 0.1 and 0.2 lbs VS/cu ft/day (1.6-3.2 kg VS/m^3/day).

The next stage is to divide this minimum size by the volume of waste produced per day to find the retention time. If this retention time falls within the range of the digester system chosen, this is all to the good and the waste does not need any further dilution or concentration. *Figure 30* shows some of the ranges of operating moisture contents with the retention times of various digesters. It is not a hard and fast rule that the more diluted a waste the shorter the retention time, but there does appear to be a general trend towards this. It is obvious, however, that for a completely mixed high-rate system the dilution should be greater than for a plug-flow digester (and also for a batch process) since wastes with a high solids contents are more difficult to mix efficiently. The process chosen should take into account that a high-rate system achieves a greater reduction in organic matter with a slightly reduced gas production, while a longer retention-time system (eg. a plug-flow) may not reduce the polluting material so much, but would have a similar or slightly higher gas production.

So, if the retention time is too short for the process the moisture content should be reduced and with it the volume. A waste of 95% moisture content is twice the volume of one of 90% moisture content and three times the volume of one of 85% moisture — the retention times are thus twice and three times longer for these moisture contents. On the other hand if the retention time is too long or the waste is too thick to be

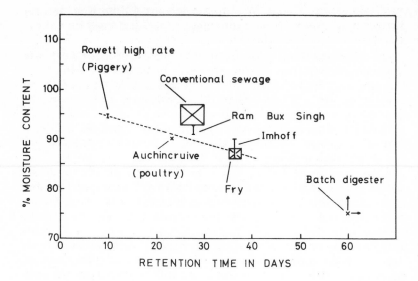

Fig. 30. Retention times of various digesters compared to their working moisture contents.

pumped or mixed efficiently, then the volume should be increased by diluting it. Thus the next stage is to find the moisture content which, when put into a digester of the correct capacity to deal with the total weight of volatile solids, will give the appropriate retention time.

There are various ways of concentrating or diluting the waste, but the first thing to do at this point is to reconsider the waste itself and how it is produced. If it is too dilute a more concentrated waste will obviously be created by using less water — this is a matter for good 'housekeeping' by making sure that no taps are left on, that drinking troughs do not overfill, that rain water does not find its way into the waste, and by considering whether less water could be used to wash down the yards, animal houses and lavatories (if human excreta is being digested). Another way to concentrate the waste is to add more organic material; this may be useful for balancing the C/N ratio correctly, but will involve re-sizing the digester. The third and most practicable way, if the other two are not feasible, is to concentrate the waste by allowing the solids to settle out and discarding the supernatant. This method leaves a highly polluting supernatant to be disposed of, which could present problems of water pollution, and public and animal health.

Dilution is perhaps easier than concentration, the main

problems being where the water is to come from and possible increased disposal costs after digestion. Water may obviously be obtained from the normal sources (mains or wells), but this could be expensive; there is no real objection to using water from the supernatant of the digested waste as a dilutant or to the use of yard washings and rain water. The main rule for dilution is that it must be carefully controlled and the correct amount of water added to the waste. Once the moisture content and retention time have been fixed, we should make sure that the nitrogen content is still below the toxic limit, especially if the waste has been concentrated.

Thus we have arrived at the size of the digester, the retention time and the moisture content required for that retention time. In order to illustrate the process of sizing of a digester further we will follow a hypothetical example through the various stages.

Our example is a farm where 1000 pigs are kept indoors on slats, ie. with no bedding material, and conventionally fed, ie. non-liquid feed. The pigs each produce one gallon per day (4.5 litres) on average and another gallon per pig is used to wash down the houses. So a total of 2000 gallons of waste is produced per day. Now for this calculation each pig excretes on average 1.6 lb of total solids daily, which at 83% volatile solids, amounts to 1.3 lbs VS per pig per day. The total volatile solids produced per day is thus 1300 lbs (585 kg). The moisture content is 92%.

Each pig excretes about 0.05 lb of nitrogen per day (0.02 kg) or 50 lbs (22 kg) N in total, so that the C/N ratio is about 5 to 1. There is therefore sufficient nitrogen available to ensure digestion of all the organic material. The concentration of nitrogen in the waste is about 2500 ppm, which is approaching the toxic limit. Dilution may have to be considered at a later stage, even though only 50% of the nitrogen will be in the ammonia form. (If there are other carbonaceous waste materials, the piggery sludge will contain enough nitrogen to help digest them if the two are combined.)

So 1300 lbs of VS per day (585 kg) will require between 13,000 to 6,500 cubic feet (364-182 m^3) at loading rates of 0.1 to 0.2 lb VS/cu ft/day (1.6-3.2 kg/m^3/day). If 0.17 lb VS/cu ft/day (2.8 kg VS/m^3/day) is taken as a suitable loading rate, the minimum volume of the digester should be 7650 cubic feet or 47,660 gallons. At 2000 gallons per day (9.1m^3/day) this gives a retention time of approximately 24 days (or a range of 20-40 days for the other loadings). This retention time is within the range of conventional and plug-flow digesters and no further dilution or concentration need be applied to the waste if this sort of system is to be used.

If a retention time of 35-40 days is required (eg. if it is decided to use Fry's system), either a bigger digester can be built, ie. one with a loading rate at the lower end of the range, or the waste may be concentrated. This latter solution is probably not very advisable, for to concentrate the waste to 87% moisture content, which would give 1200 gallons/day (5.45 m^3/day) and a retention time of 87%, would also increase the total nitrogen content to 4,167 ppm which is above the toxic limit. The only way to overcome this would be to increase the C/N ratio by adding more organic matter and hence effectively balancing the C/N ratio. In this case the digester would have to be re-sized with the new daily volatile-solids loading.

If a high-rate digester is chosen with a retention time of 10 days, the daily volume should be increased to 4765 gallons (21.6 m^3/day) which would have a moisture content of 96.6%. The nitrogen concentration here is 1050 ppm. At 96.6% moisture content the waste would be watery enough to enable efficient mixing to take place.

4. Heating and Insulating Requirements

In most cases, except perhaps in a tropical climate, the digester is heated and maintained at the temperature chosen (around 86-95° F, 30-35° C). Before we can estimate how much energy will be required, we should first find out the average and minimum winter and summer temperatures. Local meteorological offices will be able to supply this information quite easily. Secondly we should find out, or estimate, the temperature of the waste as it will enter the digester (if it is put in fresh it will probably be several degrees higher than if it has to be kept for a day or so). Now that we have decided upon a size for the digester, we can work out the surface area exposed to the air. A larger digester has less surface area per volume than a smaller one, so that heat losses through the surface will be proportionately lower. Therefore in shaping the digester, the one with the smallest surface-area to volume ratio would be best. Heat is needed to warm up the sludge coming into the digester and to compensate for the radiation losses from the digester surface. In the first case the energy required to heat up the incoming waste may be calculated from the formula:

$$\text{Energy required to heat the slurry} = \frac{\text{Volume of slurry per hour} \times}{(T_{\text{(operating)}} - T_{\text{(input)}})} \times K.$$

For BTU required = Gallons of slurry $\times (T_o - T_i)° F \times 10$.

For K calories required = Litres of slurry $\times (T_o - T_i)° C \times 1$.

For instance the heat required for 2000 gallons (9100 litres) of pig slurry to be heated daily from 41° F to 95° F (5° C to

$35°C$) — as is typical in winter — would be:

BTU = 1,080,000 BTU/day = 45,000 BTU/hour
or Kcals. = 273,000 Kcals./day = 11,375 Kcals./hour.

There are various ways of raising the temperature, and they should be considered in the light of the existing situation. The sludge can be heated either once inside the digester or as it is put in. Since some sort of heating device will probably be needed to compensate for radiation losses, the former method would require only one sort of heating system. However, it would obviously help if the input slurry was somewhat warmer, and either the warmed supernatant can be used as dilution water or the effluent can be used to heat up the influent. If the sludge has to be settled before the supernatant can be taken off, the temperature of the supernatant would probably have fallen considerably. If the supernatant is taken directly from the digester, as in conventional and displacement digesters, the warmth will be retained. If there is a separate source of warmed water, which could also be used for dilution, this should be considered as well. The temperature of the diluted wastes can be calculated by applying the formula:

$$T_{(diluted\ waste)} = \frac{(V_1 T_1 + V_2 T_2)}{(V_1 + V_2)}$$ where V_1 and V_2 are the volumes, and T_1 and T_2 are the temperature of the waste and dilution water.

The heat lost from the digester surface is slightly more difficult to calculate, for the digester will be made up of different materials which conduct the heat away at different rates. Also some parts will be in contact with the earth and some in contact with the air. However, an estimate of the heat lost through the surface of the digester can be made by supplying the formula:

$$\text{Total hourly heat loss} = \frac{A\ (T_i - T_o)}{\frac{1}{f_i} + \frac{b}{k} + \frac{1}{f_o}}$$

in which A = surface area in square feet or square metres

T_i = Temperature inside in $°F$ or $°C$

T_o = Air temperature outside in $°F$ or $°C$

f_i = surface coefficient of transmittance per $°F$ or $°C$ to inner surface

f_o = surface coefficient of transmittance per $°F$ or $°C$ from outer surface

b = wall thickness in inches or metres

k = thermal conductivity in $BTU/hr/ft^2/inch$ or Kcals./hr/sq m/m.

99

Table 20.   Thermal conductivities and transmittance coefficients of various materials. [9] [1] [2] [18]

Material	Thermal Conductivity (R)		(1/f) Reciprocal of transmittance coefficient.	
	Btu/hr/ft²/inch/°F	Kcals/hr/sqm/m/C°	Btu/ft²/°F/hr.	Kcal/cm²/°C/hr.
Wool	1.0	0.124	0.71	
Methane (saturated)	3.00	0.372		
Brickwork	4.0-5.0	0.5-0.62	0.71	
Concrete	6.7	0.83	0.71	
Steel	40-50	5-6.2	0.5	
Expanded polystyrene	0.24	0.03		
Fibreglass	0.24	0.03		
Granulated cork, slag wool	0.32	0.04		
Fibre board	0.40	0.05		
Loose asbestos, plasterboard	1.13	0.14		
Heavy damp clay	8.23	1.12		
Loam	9.11	1.13		
Sludge to Methane			0.6	0.123
Methane to underside of roof			0.6	0.123
Roof to outside air			0.25	0.051
Sludge to inside walls			Negligible	
Vertical walls to outside air			0.4	0.082

Table 20 shows some of the characteristics of digester building materials which will help us to calculate the heat loss. Thus, if a mild steel ¼" (6.5 mm) thick digester tank of about 15 foot high by 13 foot radius is used, the surface area in contact with the sludge (i.e. the walls) will be about 1200 square feet. (Heat losses through the bottom are difficult to calculate, since they depend upon the degree of saturation of the earth and the rates of flow of ground water, etc. For the purposes of these calculations they are left out.)

If the temperature is 41°F (5°C) and the inner surface temperature is 95°F (35°C) — i.e. the operating temperature — the total hourly heat loss is:

$$= \frac{1200 \times (95\text{-}41)}{0 + \frac{0.25}{46.3} + \frac{1}{2.0}} = \frac{1200 \times 54}{0.5 + 0.005} = 128{,}316 \text{ BTU/hour } (32{,}336 \text{ Kcals/hr})$$

($\frac{1}{f_i}$ is zero since the transfer of heat from the sludge to the inner surface of the digester wall is considered to be instantaneous).

If a 3" layer of insulating material such as expanded polystyrene is used. the formula may be modified to take in the extra layer; thus total hourly heat loss

$$= \frac{A\,(T_i - T_o)}{\frac{b}{k}\,(\text{for steel}) + \frac{b}{k}\,(\text{for insulation}) + \frac{1}{f_o}}$$

$$= \frac{1200 \times (95 - 41)}{\frac{0.25}{46.3} \quad \frac{3}{0.24} + \frac{1}{1.4}} = 4{,}904 \text{ BTU per hour } (1{,}236 \text{ Kcals/hr})$$

or a saving of 96% of the heat that would have been lost if no insulation had been used.

Similarly a calculation may be made for the heat loss through the roof which is not in contact with the sludge. If the area of the roof is 510 square feet, the calculation takes into account the transfer of heat from sludge to methane, methane to gasholder, and gas holder to air. Thus

$$\frac{510 \times (95\text{-}41)}{0.6 + 0.6 + 0.25} = \frac{510 \times 54}{1.45} = 18{,}993 \text{ BTU per hour } (4{,}786 \text{ Kcals/hr})$$

and if insulated

$$\frac{510 \times 54}{0.6 + 0.6 + \frac{3}{0.24} + \frac{1}{1.4}} \qquad \frac{510 \times 54}{14.4} = 1{,}913 \text{ BTU/hr } (482 \text{ Kcals/hr})$$

a saving of 90% of the heat lost if not insulated.

So for an insulated digester we only need to add about 6000 BTU/hr (1512 Kcals/hr) to compensate for heat losses from the surface of this size of digester.

Thus the total BTU requirement for this 2000 gallons per day unit on an average winter's day might be about 1,080,000 + 144,000 = 1,224,000 BTU (12¼ Therms) or 308,450 Kcals/day. The amount of gas produced by 1000 pigs per day at 6.5 cubic feet of gas of 600 BTU/cu ft per lb of volatile solids is equivalent to 5,070,000 BTU per day (1,277,640 Kcals/day). Thus the heat requirement is about one quarter of the energy produced as methane; allowing for inefficiencies in heat transfer etc., this requirement is usually taken as one third of the total energy produced.

The size of boiler or the amount of hot water (from generator cooling water etc.) which will be necessary to fulfill this heat requirement may be calculated from these figures. An allowance should, of course, be made for the inefficiency of the heat exchanger — a factor which depends upon the system chosen.

We have seen what a difference insulation makes — in the example given it would have been ridiculous to operate the digester without insulation. The thickness requirement of insulation can be calculated from the above formula, and most of the normal insulating materials are suitable. However, the insulating material must always be kept dry, for moisture reduces its efficiency in retaining the heat. This is a big disadvantage of fibreglass, for instance, compared to expanded polystyrene. The latter on the other hand may tend to trap pockets of biogas which have escaped. Other methods — such as the use of straw, covering the digester with earth, or sinking it into the ground — have been used effectively, but again the moisture content should be kept low. Some people have used compost to insulate their digesters. In this way the heat generated by the compost can be used to warm the digester as well as to insulate it, but this may be unnecessarily complicated.

Besides the digester itself, the influent and effluent pipes for the heat exchangers, and the gas pipes, should be insulated. This is not only to minimise heat losses from the pipes, but also to prevent them from freezing — a situation which could be dangerous if gas pipes were ever blocked up by ice.

Apart from insulation the only other measure which will conserve energy is to situate the digester in a sheltered place, because as the wind speed increases so does the rate of heat loss from the surface (the transmittance coefficient is increased three fold for a wind speed of 15 miles per hour).[12]

5. Shape and Dimensions — Ground Area — Choosing a Site

Once the volume of the digester has been calculated, we next have to decide upon its shape and dimensions. Most vertical digesters are cylindrical, and the smaller sizes tend to be taller than broader, for instance the farm-scale unit at the Rowett Institute is 12 feet high by 8 feet in diameter (3.65 x 2.43 m). Their estimations show that the largest size of digester which is feasible with these proportions is about 20,000 gallons (90.8 m^3), which would be 28 ft high by 11.7 ft in diameter (8.5 x 3.6 m). Tall and thin digesters certainly occupy less space, but the engineering problems become greater the taller the tank. For volumes larger than 20,000 gallons a wider, squatter vessel is more suitable, but as the diameter increases simple turbine stirrers become less efficient at mixing. The larger digesters approach the almost 'spherical' shape found in many sewage-works models. If you are planning a big digester, you will almost certainly require specialised engineering advice.

The cylindrical shape is usually chosen for a number of reasons. A cylinder has no corners or dead spots and so mixing will be more efficient; it also offers a small surface area per volume, which is important for minimising heat losses; the ground space which a cylindrical tank occupies is less than that required by a cube-shaped tank; and lastly it is a fairly easy shape to construct, whether it be made of steel, concrete or butyl rubber.

One of the most important considerations when designing the digester are the foundations. The larger the tank, the more weight of sludge and the greater the pressure upon the foundations. Obviously a digester should never be built upon unstable or marshy ground without adequate precautions. Similarly the supports for the walls must be sufficient — one of the reasons why digesters are sometimes sunk in the ground is to make use of the support offered by the surrounding earth.

Horizontal digesters are usually cylinders (or semi-cylinders) which are laid on their sides. Although they occupy more land space than the vertical type, the pressure of the sludge is spread over a larger area. This means that the foundations and wall supports do not have to be as strong. As a result the cost of construction can be considerably lower. The horizontal digester can be very large indeed; Fry quotes the use of a steel tank 100 feet long and 25 feet wide (31 x 8 m). This would have a volume of 45,200 cubic feet or 282,000 gallons (1282 m^3) and would occupy a total ground area of 2500 sq ft (233 sq meters). A vertical cylinder of this size would be a major engineering task.

At the other end of the scale the butyl-rubber pillow bags produced by Biogas Plant Ltd have capacities of 500 and 3000 gallons (2.3 m^3 and 13.7 m^3). They are rectangular in shape, being 12 ft x 6 ft (3.7 x 1.9 m) and 27 ft x 8 ft (8.4 x 2.5 m). The 3000 gallon size occupies a ground area of 216 sq ft (21 sq m) and can be directly compared to the farm-scale tank of the same volume at the Rowett Institute which takes up 50 sq ft (4.6 sq m), i.e. less than a quarter of the area.

The ground area that the digester will occupy is the first criterion which must be applied to any prospective site. As well as considering the tank, the sizes of the gas holder, holding tanks, pump and boiler house must be estimated. Spaces between the various tanks and access to them must be allowed for when deciding the total land area. The size of the gas holder is calculated by multiplying the hourly rate of gas production by the length of storage time required. Most gas holders (except the butyl pillow tank ones) will be upright cylinders, and the dimensions can be estimated in the same way as for a digester. The holding tanks are also sized from the required storage capacity, but on many farms there will already be tanks in use which will serve this purpose. If at all possible existing facilities such as tanks or lagoons should be used, and the position of these may well be the decisive factor in siting the whole system.

In practise the actual site for the digester will probably be fairly obvious, but factors such as the distance from animal houses, the lie of the land, and, ease of access for maintenance and removal of sludge, should all be taken into account. Most important of all the digester should be situated in a place which will allow room for any foreseeable increases in production.

It is an ideal time, when planning a digester, to reconsider the whole waste-collection system. Indeed a digester can not really be planned without regard to the method of transferring the waste from its source to the anaerobic tank. In this respect the lie of the land is important, for it will determine the need for pumps or scrapers in channels. In order to use the slopes to the best advantage, the digester may be sunk into the ground (see Fig. 12). If they are not fed from a manually filled header tank, however, vertical digesters above ground will almost certainly need a pumped feed, unless the unit is below the level of the animal house. In horizontal digesters the sludge requires a much smaller 'head' to displace it down the tank, and as a result the need for pumping will not be so great.

The area around a digester, however, should not be too steep, for if there are spillages of slurry, it will flow downhill and could cause pollution. For this reason digesters should not be situated on the banks of a river or lake. It is a good idea if some nearby ground is available for use in emergencies to empty the whole system quickly. One or two fields should be the minimum space between the digester and the river to cater for these rare occurrences.

For when spillages do occur, an adequate water supply is essential to clean up the digester area, although again care must be taken not to hose the spilt sludge down into the river. A good water supply is also necessary for both diluting the waste if it is too strong and for cleaning the equipment regularly.

The other essential service is of course electricity, although if the biogas is used for generating, this may not be quite so important. Nevertheless there will be occasions when the digester is out of action or being started up, when outside electricity supply is necessary to power the pumps, stirrers or even a simple light bulb. Perhaps a two-digester system could cover this eventuality, so that not 'all your eggs are in one basket'.

Finally the digester must be easily accessible in all weathers, for sludge removal, whether by wheelbarrow or tanker, has to continue all the year round. The site has to be reasonably conspicuous, so that maintenance etc. is not forgotten, but also out of the way of vandals. It should also be fairly sheltered from the wind, but not so close to animal and human habitation as to become a potential danger if gas leaks.

6. Digester Design — Check List

1. *Waste characteristics*
   - Volume produced per day.
   - Weight of volatile solids produced per day.
   - Moisture content.
   - Nitrogen content.
   - C/N ratio.

2. *Can the waste be digested as it stands?*
   - Size of digester.
   - Retention time.

3. *Does the waste need further treatment?*
   - Dilution.
   - Concentration.
   - Chemical addition, e.g. Lime
   - Addition of more organic material to alter the C/N ratio.
   - Large solids shredding or maceration.
   - Adjusted size of digester/retention time.

4. *Site for digester*
   - Is the lie of land used to the best advantage?
   - Are pumps needed; if so at what point: input to digester, sludge recirculation through heat exchanger, output from digester.
   - Is the waste pumpable, and are the pumps protected from large solids by means of a filter?
   - What happens in the event of pump breakdown? Need for stand-by pumps?

5. *Pipework for sludge*
   — at least 3″ in diameter.
   — Easily drained and cleaned out.
   — Flexible system, to allow for changes in operation.
   — Must not siphon sludge from the digester giving a negative pressure inside.
   — Insulation.

6. *Digester vessel*
   — No waste or hidden spaces.
   — Provision of portholes and manholes for sampling and access.
   — All moving parts, e.g. stirrers, recirculating sludge pumps, to be outside the digester to give ease of maintenance.
   — Facilities for scum and sediment removal.
   — Pressure relief valve for excess gas production.
   — Thermometer, thermostat.
   — Heat losses. Heat-exchanger needed/size of boiler?
   — Insulation.
   — Protected against corrosion.

7. *Gas holder and gas pipework*
   — Gasholder to maintain pressure of about 8″ water gauge.
   — Pressure relief valves.
   — Good ventilation.
   — Pipework — must have non-return valves.
   — Condensate traps.
   — Flame traps.
   — Leak proof.
   — Insulation.
   — Protected against corrosion.

8. *Safety*
   — Digester system built on firm ground — not marshy, etc.
   — Electrical fittings near digester should be flame-proofed and maintained in good condition.
   — Pumps, valves, boilers etc, if housed, should have adequate ventilation to allow gas to escape if leaks occur.
   — Earth digester, gasholder and pipework.

9. *Uses for the gas*
   — Is gas use balanced against gas production?
   — Is gas holder sized correctly to even out irregularities?
   — In the event of digester breakdown, what alternative sources of energy are available?
   — Scrubbing and compression?

10. *Uses for the sludge*
    — Is the production of sludge balanced carefully against the disposal of the sludge?

- What is the rate at which sludge can be applied for maximum utilisation of the fertiliser content by the crops?
- What happens when sludge cannot be removed, e.g. due to very wet weather conditions?
- Does the sludge need concentration by settlement and supernatant removal; if so what happens to the supernatant?
- Where can sludge in the digester be put in an emergency without creating pollution? Sacrifice land?

7. Materials

Table 21 shows some of the advantages and disadvantages of different materials which can be used for the various parts of the digester system. This is not an exhaustive list, and further information on the uses of these and other materials can be obtained from research institutes and advisory bodies, some of which are listed in the next chapter.

Insulating materials

Insulating materials work on the basic principle of preventing the air from moving to or from the insulated surface. They usually consist of finely divided fibres or particles which, being loosely packed, contain a large volume of static air. The commercially sold insulating materials such as fibreglass wool, expanded polystyrene (balls or blocks), expanded polyurethane, vermiculite and cork are long-life, resistant materials, which take good advantage of the trapped air principle. Fibreglass and, to some extent, vermiculite suffer from the disadvantage, which the others do not, of losing their insulating properties when they become wet and sodden. If glass fibre is used it should be well protected from the rain by encasing it in a plastic covering. Of the natural, cheaper insulating materials, straw, wool, and earth have been successfully employed. Earth will last indefinitely, the others not so long, for they will be attacked by rats, mould, etc. They will have to be replaced regularly, perhaps annually in the case of straw. Once again water must be prevented from seeping into the material as this not only reduces efficiency but also encourages decomposition. Water movement in and through earth is one of the major factors which influences the rate at which heat is lost from a digester buried in the ground. The support and attachment of the insulating material to the digester is an important consideration. Glass fibre, expanded polyurethane and wool have to be supported by binding them to the digester. Particulate materials such as vermiculite and polystyrene balls are good for covering horizontal surfaces, but less suitable, unless rigidly confined, for

Table 21.

Material	Use	Comments on Use	General Comments
Steel 1. Mild	Tanks (collecting/settlement) Pipework Reinforcement Digester Gas holder Fittings	Can withstand high pressure High thermal conductivity — will need good insulation	Generally very strong, durable versatile and easily obtainable. Unless treated (painted, epoxy resin coat plated, etc.) it will corrode. High cost and when in large tanks will need suitable foundations to support its weight.
2. Stainless	Specific fittings and areas where corrosion is likely, e.g. Pressure valves and controls		Will not corrode but its use is restricted owing to its high cost.
Concrete	Tanks Digester Foundation work	Either prestressed or reinforced unless of massive construction Good thermal insulation	Robust, generally strong and readily available but it can be attacked by acid conditions. Very heavy, fire resistant, needs no maintenance. Easily moulded. Additives may improve the properties, e.g. sulphate resisting.
Butyl Rubber	As a free standing 'pillow' tank (closed) Digester Gas holder Portable sludge/water container on trailer As an impermeable lining in a supporting structure (closed or open) Digester Gas holder Collecting tanks	It may need a mesh bag to contain it An ideal support is corrugated steel sheets bolted together Sharp edges, stones, etc. should be avoided	Cheap. Versatile with a long life compared with other imperme- able membranes. An ideal DIY material since it can be patched. But it damages easily. Thickness varies from 0.02" –0.08" It will need insulation if used in a digester. It expands to accommodate in- creased gas production. Not attacked by most wastes
	Lagoons	Soil conditions dictate suitability of a lagoon depending upon slope of sides (never more than 45°) and height of water table. Butyl will need securing in an anchor trench around periphery	but attacked by organic solvents. e.g. paraffin.

Table 21. (cont.).

Material	Use	Comments on Use	General Comments
Butyl Rubber (cont.)	Lay-flat piping not less than 3" OD		Not suitable for very large installations unless given adequate structural support. Impermeable to both gas and liquid. Will burn easily once ignited. Butyl may be reinforced with nylon, giving strength with flexibility.
Plastics Polythene P.V.C. sheet	General impermeable lining	Not as suitable as butyl	Cheaper than butyl, but not such long life. Plastic is more susceptible to damage.
Glass Fibre	Preformed tanks	Usually only for small tanks owing to cost.	A strong material which is completely corrosion resistant. It is versatile in operation but can be extremely costly.
PIPEWORK Steel	Widespread use for sludge and high pressure work	Large diameter for sludge (not less than 2", preferably more)	Durable and abrasion resistant. Should be protected against corrosion. Very heavy and expensive.
Copper	Gas pipes. Boiler pipework and fittings	Standard use in gas installations	Good thermal conductance for heat exchanger and scale resistant. Compression fittings, etc. make copper pipe easy to use for DIY. If connected with other metals corrosion could be caused by galvanic cell action. Expensive.
Plastic Pipe (UPVC or ABS)	All types of pipework Range from ½" diameter upwards	High pressure applications require thick walled pipe	Corrosion free, suitable for high and low pressure. Cheap and ideal for DIY work. Fittings are either glued, screwed or push fit depending upon pressure. Long pipe runs need support. Electrically non-conducting. Fragile and not so durable.
Pitch fibre ⎫ Saltglaze ⎭	Very liquid wastes only e.g. supernatant.	Usually only for drainage Unsuitable for pressure work	Cheaper to install. Expensive to install. Corrosion resistant. Only large diameter 3" +.

109

insulating vertical walls. Straw bales can be piled up easily round a digester, while cork and polystyrene could be stuck on in the form of thick tiles. Earth will have to be banked up in order to support itself around the digester.

The insulation of pipes to prevent freezing and heat loss is best achieved by long thin strips of fibreglass wool wound around the pipe. Another method would be to clip on expanded polystyrene tubes which are moulded to the shape of the pipe. Natural materials are less easy to apply and not so effective.

8. Legal Aspects and Grants

There appear to be no legal restrictions at present preventing people from building and operating anaerobic digesters to produce methane. There are, however, a number of legal requirements which have to be considered especially if a permanent digester is planned.

The first point is co-operation with the Local Authority Planning Department. For farm buildings of less than 5000 sq ft planning permission is not usually necessary, except where the scenic amenity might be impaired. For industrial purposes planning permission for waste treatment should be sought. The Local Authority will be able to advise as to whether permission is required in particular instances. The Local Authority also administers the law for controlling the public nuisance of smells and odours. The Local Environmental Health Officer can only insist on measures to control smell, e.g. anaerobic digestion, if someone in the vicinity has complained.

Building Regulations approval may also be required (again from the Local Authority), but since there are no specific regulations for anaerobic digesters, as there are for septic tanks for example, this may not be a problem. However, the Local Authority may well refer this to the technical section of the Regional Water Authority, who would in any case be better able to advise on the suitability of construction of an anaerobic digester.

The Regional Water Authority have control over the discharge of effluents, both industrial and agricultural, to water courses and to sewers. The Control of Pollution Act 1974 has brought together many of the clauses relating to water pollution which were previously dealt with under the Rivers (Prevention of Pollution) Acts 1951 and 1961, the Water Resources Act of 1963 and the Public Health Act 1936. The new act will probably become law at the end of 1975, when it is hoped that the administrative machinery (Part 2 sections 34-41) will be set up. The main provisions (Part 2 sections 31 and 32), which include the public discussion of consent applications, will come into force six months after that date. Some of the clauses may have to be omitted because their enforcement is likely to be too expensive in the current economic situation.

Agricultural wastes are treated as industrial ones from the point of view of discharge to a river or sewer, and a consent must be obtained from the Regional Water Authority to start a new discharge or to change the characteristics of an old one. Existing discharges should have consents already. When fresh consents are applied for, characteristics such as the nature and composition (i.e. BOD, pH, Suspended Solids), the temperature, the maximum daily discharge and rate of effluent discharge, will have to be stated.

A waste water, such as the supernatant from the digester when discharged to river or sewer, will require a discharge consent even if it has been treated further. The spreading of solid farm wastes on agricultural land does not necessarily require a consent, although discussions are continuing as to how to include this in the Control of Pollution Act, 1974. However, the Regional Water Authority can issue a notice to a farmer to abstain from various practices, such as the spreading of manures and sludges on his land, if they consider that these have caused or will cause pollution. They are also concerned with the run-off of nitrates and phosphates into rivers and ground water which will later be abstracted for drinking water. The digestion of sludge will probably make the placing of such a notice less likely, since much of the polluting matter has been destroyed in digestion. The Ministry of Agriculture, Fisheries and Food is preparing several Codes of Practice for prevention of pollution by farm wastes (both livestock manures and silage effluent). If it can be shown that the farmer is following these Codes of Practice, he is less liable to prosecution in the event of a pollution incident.

The third legal point to be considered when planning a digester is that of safety. The Health and Safety at Work Act 1974 was introduced to protect people at their workplace. It imposes duties on designers, manufacturers and suppliers of equipment to ensure that they are safe. This same act provides for a Health and Safety Executive to administer its various requirements through the Factory and the Farm Safety Inspectors, whose initial advice should be sought if digestion and methane production are planned. As potentially dangerous activities anaerobic production and storage of methane warrant their scrutiny, although there are no specific statutory regulations. If the methane is to be stored in ordinary gas holders at about 8-inches water gauge, the local factory inspector should be informed, and his advice sought as to the safety of the equipment. He will be even more concerned if the gas is compressed to say 2000 psi, for further safety precautions will then have to be taken. If the gas is to be liquefied (highly unlikely for small operations), the Liquid Methane Order comes into force. Liquid methane is classified as a petroleum spirit, and, as such, has to be stored under licence from the Local Authority.

Further advice about safety measures can be obtained from the local fire prevention officers. Information on materials and building techniques can be obtained from such bodies as the Fire Research Organisation, the Building Research Establishment, and the British Standards Institute. The Institute of Electrical Engineers with the B.S.I. publish codes of practice relating to the types of electrical equipment required in certain hazardous areas. (B.S.I. No CP 1003 — Electrical apparatus and associated equipment for use in explosive atmospheres of gas or vapour other than mining applications, parts 1 and 2).

Other restrictions which may be imposed are those of insurance companies, which may charge high premiums if a digester producing methane is installed. In one particular case the company insisted that the digester was situated 600 feet from the farm buildings.

In the same way as there are no specific regulations for anaerobic digesters, there are no specific grants available for their installation. The Ministry of Agriculture, Fisheries and Food offer a 20% grant for farmers and horticulturalists (Farm Capital Grants Scheme), which covers farm buildings, stand-by electrical generators and holding tanks for pollution-control systems. However, it does not cover mechanical equipment, so whether parts of a digester system would qualify for a grant or not, is a matter for advice at the local Ministry offices. There is another 10% grant, the Farm and Horticulture Development scheme, (in line with EEC regulations,) for which plant and machinery could qualify, but the project must be approved as part of recognised development. The most important form of official help towards the installation of equipment is a tax concession of 100% in the first year; this is considered to be a large enough incentive without other grants.

The Department of Energy operate an Energy Saving Loan Scheme for industry. This applies to projects costing not less than £10,000 and is available for such things as the installation or modification of plant so as to enable waste organic materials to be used as a fuel. The lending rate, as at May 1975, was 11½% per annum. A regional development grant under the Industry Act 1972 could perhaps be applied for. The grant would only be given if the digester forms part of a recognised development in specific regions (North of England, Scotland, Wales, Cornwall).

Although mention is made of these grants and laws, it is not yet clear whether they would be given for anaerobic digesters. Success in obtaining official aid will depend entirely upon the situation.

Legal requirements and grants will obviously be different in other countries, and although the position may be similar to the UK, foreign readers should make further enquiries.

9. Cost-Benefit Analysis

It is impossible to carry out a meaningful cost-benefit analysis of the anaerobic digestion system without reference to particular cases; however, set out below are some of the items which ought to be costed, and balanced against the benefits, in order to discover whether a project is viable on grounds of economy and energy.

Obviously not every item will apply in each case, for the list merely gives guidelines for the would-be methane producer. Apart from the gas and the fertiliser produced, the other benefits may be difficult to assess, perhaps even negligible in some cases. If they do pose difficulties, the best approach is probably a comparison with the cost of alternative treatment, if this is necessary.

Another uncostable benefit — if you do build a digester and run it successfully, the problems that you will have already solved will be of enormous benefit in furthering the development and spread of the process in the future.

1. Calculation of Cost Advantage:

A. *GAS:*

    a) Net annual gas production in cu. ft. (Gross production less process usage) . . . . : .

    b) BTUs per cu. ft. . . . .

    c) Energy value (a x b)   BTUs   . . . .

    d) Existing cost per unit of energy   £/BTU   . . . .

    e) Annual Cost Saving (c x d)   £ . . . .

B. *FERTILISER:*

    f) Incremental N availability (digested less existing)*   units/ton   . . . .

    g) Existing cost per unit of N   £ . . . .

    h) Fertiliser cost advantage per ton (f x g)   £ . . . .

    j) Annual tonnage produced   . . . .

    k) Annual Cost Saving (h x j)   £ . . . .

    l) Total annual tangible cost advantage (e+k)   £ . . . .

C. *INTANGIBLE COST ADVANTAGES:*
Pollution control **
Reduced odour **
Improved disease control.

Eased pasture management due to reduced souring
Possible sale of dried material

2. Calculation of Running Costs:

A. *VARIABLE COSTS:*

		Existing System £/ann.	Anaerobic Dig. £/ann.	Cost/ Saving £/ann.
i)	Energy for:			
	Collection of waste	. . . . . .	. . . .	. . . . .
	Disposal of effluent	. . . . . .	. . . .	. . . . .
ii)	Water usage	. . . . . .	. . . .	. . . . .
iii)	Labour cost of:			
	collection	. . . . . .	. . . .	. . . . .
	plant supervision	. . . . . .	. . . .	. . . . .
	spreading of solids	. . . . . .	. . . .	. . . . .
iv)	Energy for process:			
	Gas scrubbing	. . . . . .	. . . .	. . . . .
	Heating	. . . . . .	. . . .	. . . . .
	Mixing	. . . . . .	. . . .	. . . . .
	Pumping	. . . . . .	. . . .	. . . . .
	Compression	. . . . . .	. . . .	. . . . .
v)	Chemicals	. . . . . .	. . . .	. . . . .
m	Total Variable Cost/Saving per annum.			. . . . .

B SEMI-VARIABLE COSTS:

i)	Maintenance	. . . . . .	. . . .	. . . . .
ii)	Start-up energy	. . . . . .	. . . .	. . . . .
n	Total Semi-Variable Cost/ Saving per annum			. . . . .

C FIXED COSTS:
i) Depreciation:

	Life Yrs.	Gross Cost £	Net Cost £	Depreciation £/annum
Site preparation	20	. . . . . .	. . . . .	. . . .
Holding tank	15	. . . . . .	. . . . .	. . . .
Digester and ancillary equip.	15	. . . . . .	. . . . .	. . . .
Gas collection equipment	10	. . . . . .	. . . . .	. . . .
Gas utilisation equipment	10	. . . . . .	. . . . .	. . . .
Control systems	5	. . . . . .	. . . . .	. . . .

Total Depreciation per annum
ii) Cost of capital (½ net cost x cost of capital) . . . .
iii) Insurance — Increase/decrease per annum   . . . .
iv) Opportunity cost of land   . . . .
v) Increase/decrease in Rates   . . . .
vi) Grant and/or Tax Relief   . . . .

p   Total Fixed Cost/Saving per annum   . . . .

3. Total Net Cost Saving per annum:
   The total annual saving is l - (m+n+p)
   l - Total Tangible Cost Advantage p/a   . . . .
   m - Total Variable Cost p/a   . . . . . .
   n - Total Semi-Variable Cost p/a   . . . . .
   p - Total Fixed Cost p/a   . . . . .
                              Total m+n+p . . . .
                   TOTAL ANNUAL SAVING = ...

*   This also applies to Phosphate and Potash.
** A comparison with the costs of alternative treatments will be useful
   for these.

References

1. Slurry. Farmers Weekly. (Nov. 1974). 81. (21). 22.
2. Jones. Farm Waste Disposal. MAFF. (1970). Technical report 23.
3. MAFF — Farm waste disposal. (1973). Short-term leaflet 67.
4. Wheatland & Borne. Treatment of Farm Effluents. Chemistry & Industry. (1964). 357—362.
5. Fry. Methane Digesters for fuel gas and fertilisers. (1973).
6. Fry. Practical Building of Methane Power Plants. (1974).
7. MAFF. Slurry Handling — useful facts and figures.
8. Jones & Brown. ITDG Colloquium on methane generation. (1974).
9. Escritt. Sewerage and Sewage Treatment. C.R.Books, London. (1963).
10. Baines. Anaerobic Digestion of Farm Wastes. MAFF Poultry Waste Conference. (1968).
11. Mitchell. Biogas Today — a producer's manual. (1975).
12. Eckenfelder & O'Connor. Biological Waste Treatment, Pergamon Press. (1961).
13. Imhoff. Digester Gas for automobiles. Sewage works Journal. (1946). 18. 17-25.
14. Hobson. Personal Communication.
15. Biswas. 'Cow Dung Gas plant for Energy and Manure' Fertiliser News. (Sept. 1974). 19. 9.
16. Mosey. Anaerobic Biological Treatment. Symposium of Treatment of Wastes from Food and Drink Industry, Newcastle. (1974).
17. Stafford. Optimising methane production from sewage works. ITDG colloquium. (1974).
18. Biogas Plant literature.

9 How Digestion is practiced

In order to present interest in anaerobic digestion, and to help people expand their knowledge I shall describe in this chapter some of the full-scale and experimental units, be they private, commercial or research projects. As a rule most people who have had experience of methane digestion are only too happy to explain their pitfalls and successes. It is suggested that anyone who wishes to go further into the construction and running of a digester, should contact some of these people or organisations for help. Some of the leading overseas exponents of the art are also mentioned for the benefit of those who live nearer them than the U.K. There are undoubtedly many other people, both here and abroad, whom I have not mentioned. Their exclusion is not deliberate and in no way casts doubts on their findings or processes. The reasons for their omission are lack of space and acquaintance.

1. Research programmes

a). Rowett Research Institute and North of Scotland College of Agriculture. Bucksburn, Aberdeen.

This is perhaps the longest-standing current research programme which is investigating the anaerobic digestion of farm wastes. Since the late 1960s when they took over such investigations from the West of Scotland College at Auchincruive they have been examining in particular the bacteriology of digestion of piggery wastes. Starting with small laboratory-scale digesters, they have tried to identify the bacteria in the process and the various roles they play. This has developed into an attempt to find parameters for optimising the conditions for digestion.

Using a pilot scale, 100 litre, digester, which has been running for three years, they have varied parameters such as temperature (30-35 °C) and retention time (5-15 days), and have evaluated the changes in the process efficiency. They have used the results of this pilot-scale plant to design a high-rate farm-scale digester of 3000 gallons (equivalent to waste from about 300 pigs).

Photograph 1. Farm-scale digester unit of Rowett Research Institute.

After initial problems with scum build-up, which has been overcome by the installation of stirrers, the digester has been running successfully for 6-9 months. The results from this confirm the predictions made from the pilot-scale results.

The plan of their digester system is shown earlier in this book, *(Fig. 16)* and *photograph 1* gives some idea of the relative size of the unit. Estimates of costs for a similar, but simpler version of this digester unit are in the range of £4,000-5,000. For a 1,000-pig plant, costing in the region of £8,500 (1974 prices), the running costs, including amortization of capital and interest on a loan, would be about £2,000 per year. However, set against this is the fuel oil equivalent of the net gas produced, which amounts to about 70% of these costs, and the value of the fertiliser (1½ times the value of the gas).

The findings of the Rowett Research Institute have been published extensively, and they offer advice to anyone who is genuinely interested in anaerobic digestion. The enquiries have reached such a volume that they have brought out a number of information sheets which describe their work and the construction of their farm-scale digester in particular.

b). University of Strathclyde. Glasgow, Scotland.

The Department of Applied Microbiology at Strathclyde University has received a grant from the Wolfson Foundation to look into energy-systems for handling organic wastes. They are investigating the microbiology of these complex mixed fermentations in order to provide design data based upon the population dynamics of the microbes. The extrapolation of data from anaerobic sewage-works digesters has led to suggestions that one or two different temperatures should be used; that the methane yields found are the best that can be expected; and that the process has to be run on a large scale to be worth the capital investment. The Strathclyde team will be testing the validity of these assumptions on a pilot plant in Perthshire, which is at present under construction. This will provide data on the conversion of cellulose to methane under farm conditions. They hope to be able to develop models of the anaerobic systems to obtain design data for client's particular needs without much further experimentation. They envisage using small to medium sized, easily constructed, digesters which will be set going with the help of the appropriate starter culture grown in Strathclyde. They have strong links with a commercial firm, Natural Energy Systems Ltd., who will be discussed later.

c). Glamorgan Polytechnic and University College. Cardiff, South Wales.

The Department of Mechanical and Production Engineering at Glamorgan Polytechnic is investigating the various ways of improving the engineering aspects of anaerobic digestion. The Department of Microbiology at University College, Cardiff, is collaborating with them by investigating the microbiological aspects of their small pilot-scale digester. They feel that the microbiological viability of the process has been proven and that it is now time for the engineers to improve what has been a rather 'hit or miss' process and to assess its economical and commercial viability.

After a year's preparation they have designed and installed a digester of 1 m^3 capacity treating poultry waste, which was set in operation in February 1975 *(Photograph 2)*. This has been designed so that many of the parameters, such as feeding rates, degree of mixing, gas recirculation, temperature, organic loading and retention time, can be varied independently of the other parameters. The main aspects which they are studying are the charging and discharging of sludge to and from the digester, the loading and mixing techniques, gas diffusion to increase methane production and alternative solutions to problems of gas storage and sludge disposal. The underlying

118

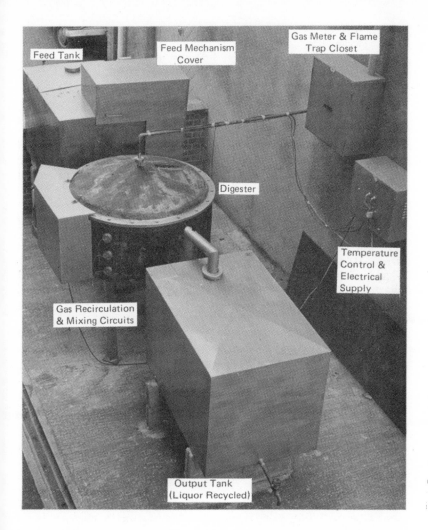

Photograph 2. Pilot-scale digester of Glamorgan Polytechnic.

aim of their programme is to design a modular system of waste disposal for automatic and continuous loading which will be easy to operate and simple to maintain. They also emphasise the safety aspect as being of prime importance.

Although they are funded through the Polytechnic, so much interest has been aroused in farmers and industrialists, that they could obtain outside cash, and even act as design consultants in due course. University College, Cardiff, as well as monitoring the microbiological and pollution-control side of the experimental digester, has carried out surveys investigating sewage-

119

works digesters and correlating operating conditions with gas production. This is aimed at optimising the methane production so that the energy produced may be used locally in factories, schools, transport systems or steel works, instead of only on the sewage works.

d). Reading University. Berkshire.

The Department of Engineering and Cybernetics have been involved in an experimental programme with both batch and continuous-flow digesters under carefully controlled laboratory conditions. Batch digesters of 15-litre capacity and small horizontal continuous-flow digesters have been run on dung from pigs and dairy cattle from the university farm (the animals have strictly controlled diets). Artificial dung has also been used so that better control of the feedstock can be achieved, especially for tests upon the effects of toxic materials. Particular aspects which they have been considering are the C/N ratio, and a simple, cheap method for measuring the carbon content of wastes. Other parameters which are measured include the gas evolved (both quantities and quality), and the overall heat balance. They consider that because the production of methane gives off a small amount of heat, this may be significant in the total energy balance of the system. The aim of their experiments is to confirm the optimum parameters and to evaluate the information for design criteria.

Photograph 3. Biogas digester sold to Manchester Polytechnic by Christopher Howarth.

e). Manchester Polytechnic.

The Departments of Physics and Biology are collaborating on a project which links up with the Intermediate Technology Development Group's methane generation panel. After initial experiments with domestic waste, other organic wastes will be investigated so as to gather as much information as possible about the requirements of the micro-organisms. At first small-scale laboratory digesters will be used to study both the biological and physical parameters such as the use of mutant strains of bacteria, temperature, pH, viscosity and gas concentrations. The next stage will be carried out on a pilot-scale plant which was bought from Mr. Howarth of Haxey (see later). The plant is shown in *Photograph 3*. The overall aim of this project is to improve the efficiency of the methane digester system in order to make it more attractive commercially, and this will be done by adjusting each of the parameters in turn and studying the effects on gas production. A detailed analysis of energy conservation within the system will be attempted in order to assess its significance. The project, which started in 1975, will probably take two or three years to complete.

f). Queen Elizabeth College, London.

The Department of Microbiology has been conducting some theoretical studies on the production of methane from pure compounds, such as carbohydrates and simple organic acids, for about two years. They are trying to establish the quantitative and kinetic aspects of the digestion of these chemicals. Their approach differs from many others in that although the bacteria came from natural sources, such as sewage-works digesters, the feedstock for the bacteria is a pure chemical. In this way one compound can be tested at a time and the microbiology of its digestion can be followed more easily. This is an ongoing project which will fill in many of the theoretical unknowns in anaerobic digestion and which may help to explain some of the many control problems which arise in practice.

g). York University.

The Department of Biology will be starting experiments in October 1975 on the 'optimisation of sewage sludge digestion'. Funded by the Wolfson Foundation, they will be collaborating with Simon-Hartley Ltd., (manufacturers of sewage-works machinery) and the Institute of Social and Economic Research at York. They intend to study the productivity of the process from scientific and economic standpoints, and there will be greater emphasis on products other than methane.

121

h). Water Research Centre. Stevenage, Herts.

The Water Research Centre at Stevenage, formerly the Water Pollution Research Laboratory, has been looking into the digestion of sewage sludges for many years. Recently the emphasis has changed from the investigation of materials which inhibit digestion, such as heavy metals, detergents and hydrocarbons, to the examination of digesting strong organic wastes from both agriculture and industry. They have over 20 one-litre batch digesters maintained at 35 °C with retention times ranging between 5-40 days. Screened and unscreened slurries from pigs and cattle have been tested, as has the digestion of pig waste, both neat and diluted, with an equal volume of water. With undiluted pig slurry the relative production of gas has been examined to find the degree of inhibition due to high ammonia concentrations.

As far as industrial wastes are concerned, they conduct treatability tests on a repayment basis and have completed trials on a strong organic waste with a high sulphide content (300 mg/l) which might have been inhibitory to the process. The gas produced contained about 0.5% (by volume) hydrogen sulphide and as a result was extremely smelly. They have also completed tests using an anaerobic filter to treat a starch waste with a low solids content and a C.O.D. of several thousand parts per million. This system provides a long retention time for the bacteria treating the waste, but a short one for the liquid (c. 1 day).

The main emphasis apart from small-scale trials is in the design of full-scale digesters. Because the design of systems for farm wastes is more critical than for sewage sludge, the laboratory's experience over many years has been utilised to design plants suitable for the farming situation. The emphasis is, understandably, more on the reduction of polluting material in the wastes than in the production of methane.

Amongst other services which the centre offers is advice on the toxicity and biodegradability of many different compounds. If a waste is known to contain one or more chemicals which might inhibit the treatment process, the WRC will be able to give information as to whether to expect problems from those compounds.

2. Private Digesters

a). Mr. Howarth of Haxey, Lincolnshire

Christopher Howarth has sold his original two-chambered batch digester to the Manchester Polytechnic and is planning to build a unit five times the size. His first trial unit consists of two heavy gauge cylindrical tanks of 200 gallons capacity each. The tanks are double skinned and insulated with 3″ of polyurethane and the sludge is headed by two immersion heaters to 82 °F. (this temperature being found to be the most economic). Although it is a batch process fed by material of 80% moisture content consisting of chicken manure and vegetable trimmings, the solids are kept in suspension by gas recirculation.

The gas that is produced (a minimum of 4.8 cu. ft./lb of waste) is passed through a wet filter and the moisture is allowed to condense out in a drier tank. The gas is compressed and stored at a pressure of 15 psi. This pilot-scale unit produces 1 therm per day (i.e. about 17 cu. ft.) Using the gas from his full-scale unit Mr. Howarth intends to heat the three sheds where the vegetables grown on his 160-acre farm are processed.

The biogas process was developed by Mr Howarth and an engineer, Martin Judd, and they have prepared a booklet describing some aspects of methane production. He will also connect those interested in buying a digester with subcontractors who will construct one. Furthermore he offers a consultancy service under the name of Biogas Ltd. (not to be confused with Biogas Plant Ltd.).

b).   Cherry Valley Duck Farm, Lincolnshire.

Late in 1974 this enormous duck farm with between half a million and a million ducks considered the possibility of using their waste to produce methane. They wanted to use the gas to heat the 24 duck houses which are at present heated by propane (burnt at the rate of 7,000 cubic feet per hour). Since nobody could furnish any reliable figures for the digestion of duck wastes, they decided to build their own trial unit of 250 gallons capacity (vertical 5ft high by 40 inches in diameter). They used this over a period of six months both as a batch process and as a continuous process feeding in 25-gallon lots at regular intervals. They found that on average about 7 cubic feet of gas was produced per pound of volatile solids. However, after the digester vessel burst under pressure (hence the need for pressure-relief valves), they discontinued the project. They feel that since they have convenient reasonably priced supplies of propane, they will continue to use this. Unless the price of propane rises dramatically they will not change over to methane, even though on purely theoretical calculations the methane evolved from the 20,000 gallons of settled solids produced per hour would be more than sufficient to meet their heating requirements. They decided also that because they were able to cope satisfactorily with their effluent at current production, the incentive of reducing the organic content by anaerobic digestion was not enough to warrant its use on their situation.

c).   Mr. Ashwell of Winkfield, Berks.

Mr. Ashwell has been trying to complete his full-scale digester for some time, but he has been held up by delays in obtaining equipment such as the pumps and heat exchangers. He plans to convert an existing 90,000 gallon glass-lined slurry tank into a digester that will treat the waste from 1000 pigs or more. This tank will be insulated with straw. At an additional cost of £6-7000 he is installing all the extra equipment, including gas recirculation and sludge pumps, and estimates that he will get his money back in 3-4 years. He hopes to be able to produce about 16,000 cubic feet of gas per day. He will use the gas to replace the butane which at present heats his piggeries. Any surplus gas will be used around the house. For him, however, one of the most important benefits will be the control of smells from the piggeries.

Before attempting this large-scale digester he conducted pilot tests over a period of two years using a 40-gallon digester built for him by Surrey University. Running continuously on mixed farm wastes from cattle, horses, pigs and sheep and being dosed at 1 gallon per day of fresh material he was able to produce about 5 cubic feet of gas per day.

3.   Commercial Organisations

a).   Biogas Plant Ltd., Easebourne Lane, Midhurst, Sussex.

Biogas Plant Ltd. are the only company in this country which sells 'off the peg' digesters. They have three systems: System I, a small dustbin-type batch digester for experimentation *(see Fig. 9);* System II, a 500-gallon butyl-rubber digester; and System III, a 3000-gallon butyl rubber digester *(see Photograph 4).* The first system with 7½ gallon sludge capacity consists of the digester and gas holder in one unit together with a manually operated agitator, an electric immersion heater (150 watt) and gas outlet pipe. The whole unit costs about £33 and is suitable for school demonstration purposes and for assessing the gas potential of particular wastes. Since the total gas produced will be only about 50-60 cubic feet over a period of two months, not much should be expected of it. The gas holder has a capacity of 50 litres or 1.75 cu. ft. which, if filled with biogas, is equivalent to 1050 BTU, or enough heat to boil four or five pints of water. Nevertheless the gas is still explosive and care must be taken even with this size of digester.

The Systems II and III will be of more interest to those who have small but regular quantities of waste to digest. The 500-

"Methane: Planning a Digester" is the first in a series of books to be published jointly by Prism Press and CTT. The second in the series will be "Heat Pumps" by John Sumner, available from June '76. In preparation are books on Windpower, Small Scale Water power and Eco-houses. Ian Hogan is busy working on what promises to be an excellent book, giving a broad view of natural energy systems, to be published in 1976, entitled "Everything under the Sun".

A special CTT paperback edition of the classic book by E. W. Golding "The Generation of Electricity by Wind Power" will be available from April '76. For details of this series, plus a comprehensive list of other books on alternative energy sources and self-sufficiency—please send us a stamped addressed envelope.

Also available through CTT are wind generators, solar collectors, heat pumps, water turbines, grain grinders, and other natural energy tools. Discounts on many of the above items are available through subscription to the CTT Association. Subscribers also receive a quarterly newsletter which reports on the latest developments in harnessing natural energy sources.

CTT
Conservation
Tools & Technology Ltd.

143 MAPLE ROAD, SURBITON, SURREY KT6 4BH
01-549 5888

Photograph 4. Biogas Plant Ltd. Butyl-rubber digester, System II.

125

gallon unit (12ft. x 6ft. in area) is the size for a small horticulturalist or an amateur with a large garden, since it could produce enough gas for domestic cooking or heating a small glasshouse. The 3000-gallon unit is of interest to the commercial operator; if he has more waste than one unit can manage, several units can be run in parallel. The two systems are similar in principle, being based largely on Fry's designs and calculations. They consist of horizontal, butyl-rubber, pillow-tanks fed with slurry from a smaller rigid reception tank. The digester tank is insulated and has an external heat exchanger on which the tank is laid. Very little site preparation is required. The 500-gallon unit costs about £400-600 depending upon how much extra equipment is purchased, e.g. gas storage bags, scrubbing equipment, water heaters and pumps.

The System III unit costs about £1200-1500 depending upon extra equipment; this compares very favourably with the estimated cost for the high-rate unit at the Rowett Institute, which is of a similar size. The two processes are not strictly comparable simply on grounds of cost, however, for the design principles are different for high-rate and displacement type digesters.

As well as selling digesters and the ancilliary equipment the director of Biogas Plant Ltd., J.C.V. Mitchell, has produced a booklet selling at £1.00 called 'Biogas Today — a methane producer's handbook'. He also sells a number of other books on methane production.

b). Biomechanics Ltd., Smardon, Ashford, Kent.

Biomechanics Ltd. are a small, specialist firm aiming to solve effluent treatment problems. After three years of intensive research and development, they have come up with their Bioenergy system *(Fig. 18)*, a form of anaerobic contact process which involves sludge recirculation in order to maintain the solids content at the correct levels.

They have tested this process at two large-scale experimental units — one treating the waste from a wheat starch/gluten/ dextrose factory and the other a piggery waste. *(Photograph 5)*. They are thus equally qualified to advise on both farm and industrial wastes. The Bioenergy process allows the bacteria to be held in the digester for at least 20-30 days but lets the liquid flow through it in 1-5 days, depending upon the strength of the raw effluent. Using their bioenergy separator the process can be tailored to treat effluents with BOD's ranging from 1500 to 50000 mg/l. and reduce them by 85 to 95%.

In the trials on piggery waste they have tested the slurry from pigs on slats and on straw. Their trial digester was able to produce about 8 cubic feet of gas per pound of volatile solids added.

126

Photograph 5.   Biomechanics Ltd. Farm-scale, high-rate digester under construction.

Although they see the Bioenergy process as being primarily for industrial waste treatment, they have estimated the costs and returns for digesters treating the waste from different numbers of pigs. They have calculated that for 1000 pigs on slats a 40,000-gallon digester will be needed at a total cost of £11,400. Since they generate electricity from the methane, the energy output is estimated at 15 K watts. From this annual operating profit, taking into account labour charges (2 hours/week) and depreciation etc., is £360.

This profit increases rapidly as the number of pigs increases, e.g. a 5000-pig unit (with a 200,000-gallon tank producing 74 K watts. at a cost of £23,000) will make an annual profit of £7,600. A 10,000 pig unit (400,000 gallon tank, 148 K watts costing £52,710) will make £13,470 p.a.

c). D.A. Knox, Staddlestones, Penton Mewsey, Andover, Hampshire. Representative of L.J. Fry, California.

L.J. Fry is one of the leading exponents of the art of running digesters in practical situations. He started in the 1950's in South Africa where he first experimented with batch and horizontal displacement digesters producing methane from pig wastes *(Photograph 6)*. Since then he has moved his operations to California and has built a number of digesters in the United States over the past ten years. For the past two years he has been operating a unit in Santa Barbara, California, under the auspices of the New Alchemy Institute.

L.J. Fry has written a number of papers and published two books, 'The Practical Building of Methane Power Plants for Rural Energy Independence' and 'Methane Digesters for Fuel Gas and Fertilisers'. The first follows the development of his system from its beginnings, outlining pitfalls and hazards. It is full of gems of practical wisdom, which can be applied to any digester system and should be read by anyone contemplating building a digester. The second book outlines the requisite essentials for an understanding of the methane digestion process and includes advice about how to build a batch digester from old fuel drums and a small continuous digester from the inner tube of a tyre.

Fry's nephew, D.A. Knox, is at present living in Great Britain and acting as his representative. In the early 1960's L.J. Fry had collaborated with the firm, Wright Rain Ltd. and designed a digester for farmers in the UK. This attempt to market his designs failed to arouse much interest at a time when fuel was cheap and plentiful. Now that fuel is more expensive, D.A. Knox is promoting his uncle's interests with an initial experiment at an electronics factory near St. Ives.

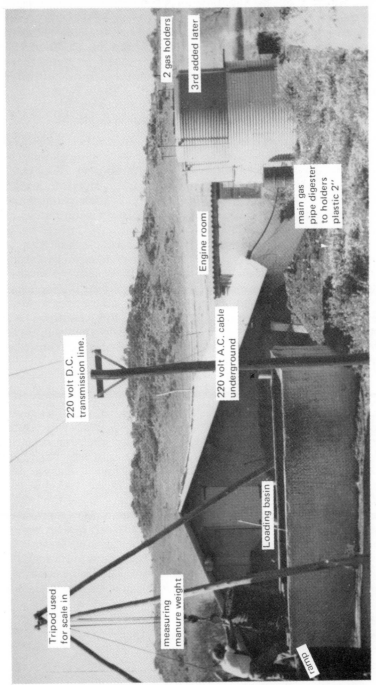

Photograph 6.   One of L.J. Fry's original South African digesters.

- 2 gas holders
- 3rd added later
- Engine room
- main gas pipe digester to holders plastic 2"
- 220 volt D.C. transmission line.
- 220 volt A.C. cable underground
- Tripod used for scale in
- measuring manure weight
- Loading basin
- ramp

129

Pig waste from a nearby farm is being used to produce methane in one of the original digesters designed by Fry. The methane will be used to provide all the heating and electrical requirements of the factory so that it can be entirely self-sufficient. Once this unit is tested and proven, Fry and Knox will consider the manufacture of similar units for world-wide sale.

d). Natural Energy Systems Ltd. Edinburgh, Scotland.

Working in collaboration with the group at Strathclyde University, Natural Energy Systems Ltd. act as a consultancy to farmers who want to use alternative energy sources. They cater mainly for the more industrialised farms, piggeries, broiler houses and hatcheries, since these have the waste production, the capital, and the staff to run a reasonably economic unit (costing in the range of £20-50,000).

They have already supplied a Scottish beef farmer with an experimental plant handling 200 tons of cow manure and straw per annum at a cost of £10,000, and they have plans for another plant to treat 350 tons of broiler litter per year. After some further development work they hope to be able to scale down the process to smaller 'packaged-units'. In the meantime they run a planning, advisory and design service to enable them to pursue this development work with reference to practical situations.

e). Sewage digester manufacturers.

There are several manufacturers who build a wide range of sewage treatment equipment including primary digesters. Secondary digesters are usually designed by civil engineers in charge of a complete project. Sometimes the civil engineers will design the primary digester as well and only use the ancillary equipment, such as mixing systems, heat exchangers and floating roofs, sold by the manufacturers. Among the better known firms making digesters and equipment are Dorr-Oliver Ltd., Ames Crosta Mills Ltd., and EIMCO Ltd. *Photograph 7* shows a typical municipal sewage-works digester system. It consists of two primary digesters and six secondary ones, and is designed to treat the sludge from about 120,000 people. In the foreground can be seen the scum layer on the top of a secondary digester.
Dorr-Oliver Ltd. — Norfolk House, Wellesley Road, Croydon.
Ames Crosta Mills Ltd. — Heywood, Lancs.
EIMCO Ltd. — St. Neots, Hunts.

Photograph 7. A municipal sewage works digester, showing scum formation on the top of the secondary digester (foreground). (By kind permission of O.C. Collins Esq. Divisional Manager, Essex Sewage Division of the Anglian Water Authority).

4. Overseas

a). Gobar Gas Institute, India.

The Indian Agricultural Research Institute at New Delhi first began their investigations into methane production from cow dung in 1939. These led in due course to the setting-up of the Gobar Gas Institute in the State of Uttar Pradesh in 1957. Through its work, and in particular that of the director Ram Bux Singh, the designs for small to medium digesters (appropriate to the Indian situation) have been much publicised. Examples of two of these designs are shown in *Figures 12 and 25*. Altogether about 9,000 digesters have now been installed in India, but even so this only scratches the surface. Ram Bux Singh and others have described the work of the Gobar Gas Institute in several publications, but they are difficult to obtain in the UK.

The smallest village-model gas plant (100 cu. ft. per day) costs about 1,500 Rupees (equivalent to £90), and in India this is considered to be one of the major drawbacks. However, the plant itself is much more labour intensive than most people in the 'developed' countries would care to operate, but its success shows that it is an ideal design for the conditions in which it has to work. It could, of course, be modified for Western conditions, but unless these change radically, the demand for a basic, time-consuming digester of this type will never be all that great.

The Gobar Gas plants replace the old-fashioned use of dried cow dung as a fuel, with a cleaner and easier gas for cooking, which also retains fertiliser value of the sludge. Estimates of the annual quantity of dung produced per year in India vary from 200-1,000 million tons, and in this context an idea of Alan Hughes in London could be exploited in order to collect part of it. His suggestion is that an old fuel tanker should be converted into a mobile digester (5,500 gallons). Each morning it would set off to collect the dung from different villages, and since it would contain a seed of digesting sludge already, the fresh dung would start to digest immediately. Gas would be produced for the tanker to run on, and any excess could be collected in an inflatable bag above the tank. The sludge would be discharged into a permanent digester on returning home in the evening and a little would be retained for seeding the next day's load. This ingenious system allows for an almost free collection service and could well be applied in areas where there are small quantities of widely dispersed digestable material.

b). T.H. Hutchinson. Tunnel Co. Ltd., Fort Ternan, Kenya.

T.H. Hutchinson began work on methane in Kenya in 1954, and after initial experiments, built a batch digester producing 100 cu. ft. of biogas per day from pig manure. In 1955 he built a larger batch plant in stone which produces 400 cubic feet per day, is still running and provides most of the energy needs for his domestic purposes, e.g. cooking, electric lighting, refrigeration, hot water. He maintains that the greatest benefit is the fertiliser value of the digested residue, which he applies to his coffee plantations.

He started marketing kits in 1957 and now sells four different types of plant. The Hutchinson methane plants have come back into their own with the energy crisis and there are now over 100 installations operating in Kenya. There are two small ready-made plants, Marks III and IV, which are operated on a continuous and batch process respectively. The Mark III plant comes in various sizes capable of treating the waste from 5-20 cows and producing 25-100 cubic feet of gas per day. The plants are supplied with piping, one light and a two-burner cooker, and cost between Shs. 950-2,300 (£1 = Shs. 17). The Mark IV can be operated with the Mark III system (using its seed) and produces 30 cubic feet of gas per day from most solid wastes, such as grass, straw, coffee pulp and sisal. The Marks I and II are larger batch and continuous digesters supplied in kit form complete with plans. They are built of stone and the Mark I plant consists of three or more separate compartments of 300 cubic feet each. These are recharged every eight weeks in rotation and yield 75-100 cubic feet of gas per day. They cost Shs. 3,000-3,500. The Mark II plant varies in size from the domestic unit for 50 cows producing 450 cubic feet of gas per day to the unit for 200 cows yielding 2,000 cubic feet of gas per day. The price varies between Shs. 3,000 and Shs. 6,000 for kit and plans.

T.H. Hutchinson also sells the plans, kit-building and operating instructions separately, especially for people outside Kenya, since export of kits is complicated. He has also sold a manu-facturing licence for India.

c). J. Coulthard. Bendigo, Victoria, Australia.

J. Coulthard bought his first digester from T.H. Hutchinson in Kenya and since then he has developed his own system in Australia. He recently won an Australian award for his invention, which he describes as the cheapest and most efficient process for producing biogas from farm waste. His system, called the fertilised chlorella process, is a methane-algal one which processes the algae grown on the digested sludge into a protein-rich animal feed.

Mr. Coulthard originally found that use of a concrete digester to treat his pig waste was uneconomic. It remained so until he developed a prefabricated metal tank lined with a butyl rubber bag at a cost of about $A 6,000. He has converted his car to run on methane for about $A 500, and claims that the cost of biogas is less than 1 cent for the equivalent of one litre of petrol (4 cents per gallon).

He has formed a company, Sanamatic Tanks Proprietry Ltd., and is planning to manufacture similar gas-producing units for sale in Australia and elsewhere.

5. Other Organisations

a). Intermediate Technology Development Group Ltd., Parnell House, 25 Wilton Road, London SW1.

I.T.D.G. was set up as a charity in 1966 by a group of engineers, economists and others, with experience of developing countries, who had the intention of investigating ways in which appropriate, intermediate technologies could be established in the third world. I.T.D.G. aim to stimulate the productivity of poor communities at the same time as creating more jobs within their existing social structures.

One of I.T.D.G.'s main tasks is to collect what little information there is and disseminate it. One way in which they do this is to bring together panels of experts, publish their findings, and recommend that they be incorporated into training programmes. They have a power panel, amongst others, looking into simple energy resources. As a result they held a colloquium in December 1974 with the aim of gathering as many experts on methane and anaerobic digestion as possible. The various papers which were presented at this colloquium are being published shortly. Currently the methane group of the power panel is compiling a bibliography of all published information. In addition to their specific work on methane I.T.D.G. publish a quarterly journal, Appropriate Technology, which contains some fascinating examples of how these ideas are applied in practice.

b). Conservation Tools and Technology Ltd., 143 Maple Road, Surbiton, Surrey.

CTT grew out of its predecessor, Low Impact Technology, early in 1975. It acts as a sort of counterpart to I.T.D.G. for Western countries, and the two are closely linked. CTT is especially concerned with energy conservation and the use of alternative resources such as solar energy, wind and water power. More importantly CTT is always actively searching for new ideas and equipment to publicise and sell. This book is being published jointly by Prism Press and CTT, and forms part of a series of books and information sheets, designed to make this sort of technology comprehensible to the layman.

CTT currently market wind generators and solar panels, and acts as agent for both Biomechanics Ltd. and Biogas Plant Ltd. for methane digesters. It also acts as a consultant on all aspects of energy conservation and advises clients about how to minimise heat losses and maximise their resources.

It has recently formed the CTT Association which acts as a 'clearing house' for information and ideas on alternative technology, and it is in contact with almost all the people working in this field throughout the world. The CTT Association issues a quarterly newsletter to keep its members abreast of recent developments.

c). Soil Association, Walnut Tree Manor, Haughley, Suffolk.

The Soil Association are not directly involved in the production of methane from organic wastes, but as a result of their interest in the use of organic fertilisers, they contribute to the methane panel organised by I.T.D.G. They, like many others, are keeping a watching brief on developments in this field. They will be able to answer most enquiries about the use of both digested and undigested sludge as a fertiliser. They also publish a quarterly newsletter.

d). Henry Doubleday Research Association, 20 Convent Lane, Bocking, Braintree, Essex.

The Henry Doubleday Research Association have produced a small booklet 'Fuel Plus Fertility' which urges greater exploitation of methane. Apart from publishing a design for a small 10-litre digester for use in schools, they are encouraging their members from all over the world to search for seed bacteria in black pond muds, etc. (especially in tropical countries). They have asked three university bacteriology departments to culture these bacteria if useful, and try them on awkward wastes such as wood shavings and newspaper. They also maintain a register of experienced people who will help those who wish to install plants.

e). National Farmers Union, Agriculture House, Knightsbridge, London SW1.

The NFU are also watching developments. Because they are able to maintain a close interest, both at an official level in research programmes and at the individual farmer's level, they have a very good idea of the 'state of the art'. Even if they cannot themselves satisfy all enquiries they will certainly be able to recommend advisers. They will also be useful to contact for other information concerning agricultural wastes, both legal and technical.

f). Ministry of Agriculture, Fisheries and Food, Horseferry Road, London SW1.

The Ministry of Agriculture, Fisheries and Food, through their various departments and branches, have perhaps the most comprehensive view of what is going on in the U.K. as far as the agricultural production of methane is concerned. The most important department for this is the Farm Waste Unit, a part of the Agricultural Development Advisory Service (ADAS), at Coley Park near Reading. They are extremely helpful and well informed and will be able to advise on all aspects of farm waste treatment: 1. legal — they are publishing Codes of Practice for the avoidance of pollution by farm wastes and 2. technical — through their own work at the various experimental farms and their knowledge of the work of others they can answer questions on most of the different types of waste treatment (both aerobic and anaerobic). MAFF also have information on the fertiliser contents and optimum rates of application for farm wastes.

g). Tropical Products Institute, Industrial Development Division, Culham, Abingdon, Berks.

The main interest of the TPI in methane production is obviously for overseas application. Although they currently have no practical programmes examining anaerobic digestion, they are carrying out a literature survey in order to have the information ready when they do start experiments.

h). Other advisory bodies, etc.

For further information on various subjects, the following organisations may be able to help:

National Institute of Agricultural Engineering, Silsoe, Bedfordshire.
Farm Buildings Association, Stoneleigh, Warwickshire.
The Gas Council, 139 Tottenham Court Road, London W1.
British Standards Institute, 2 Park Street, London W1.
Fire Research Association, Station Road, Boreham Wood, London.
Cement and Concrete Association, 52 Grosvenor Gardens, London S.W.1.
Building Research Station, Garston, Watford.
Timber Research and Development Association, High Wycombe, Bucks.

Institution of Electrical Engineers, Savoy Pláce, London WC2.
Institution of Corrosion Technology, 14 Belgrave Square, London SW1.
Institution of Gas Engineers, 17 Grosvenor Crescent, London SW1.
National Centre for Alternative Technology, Llwyngwern Quarry, Pantperthog, Machynlleth, Powys, Wales.

Further names, addresses and references can be found in the book 'Methane: Fuel of the Future' by Bell, Boulter, Dunlop and Keiller, Prism Press 1975.

10 Future Developments and Conclusions

Up to now we have been considering the art of anaerobic digestion as it is practised at present. But what of the future? How will the process and its various uses develop? What needs to be done to perfect this method of obtaining energy for when our resources are even more limited than they are today?

The interest in the process is there, both officially and unofficially, and the experimental programmes are there. Their aim is to find the optimum conditions for digestion, and some are working on the more academic features which will bear fruit in the distant future while others are intent upon solving the more immediate problems of refining and improving existing systems.

In the distant future we may have specialised bacteria to break down certain wastes in preparation for the methane producers (which are themselves highly specialised). At present we have no option but to use the 'seed' of digested sewage sludge to provide a mixture of less specialised bacteria. It will only be possible to match the bacteria to a particular waste when we have a number of digesters working on different effluents from which to choose the seed. For the treatment of agricultural wastes a sewage sludge seed will probably always be adequate, but industrial wastes may benefit from a very different set of bacteria.

Another dream for the future is the cultivation of bacteria which produce methane at a viable rate under cold conditions. This will conserve the energy at present needed to heat the digester. Cold methane-producers are most likely to be found in the muds of rivers, lakes and marshes. A similar development could be made in the separation of the acid-formers and the methane-producers into two tanks. Although they would still be interconnected, they would be independent of each other and the whole system could work more efficiently.

Some of the more immediate developments lie in the uses to which anaerobic digestion can be put. The process should never be considered in isolation, and some of the combinations with other alternative energy sources are particularly interesting. One such combination is the culture of algae to form part of the sludge fed into the digester. The sun's energy, which has promoted the algal growth, is therefore indirectly converted into a more useful form — methane. This method eliminates the intermediate stage of 'animal eating plant'. Similarly, using solar panels to heat the digester makes the best use of the sun's energy.

Although agricultural wastes offer the largest source of methane, it is the digestion of industrial wastes which could prove the most exciting. Because the latter generally contain high concentrations of only a few compounds (c.f. agricultural wastes or sewage which have very variable compositions), a 'purer' sludge can be produced. It may then be possible to abstract compounds such as vitamins and antibiotics from specialised bacteria in this purer sludge. In addition the methane produced could either be used as a fuel in the normal way or it could be converted to industrial alcohol, methanol. This can be used as a solvent or as a basic material for organic chemicals such as plastics, etc.

The most widespread use of anaerobic digestion will probably continue to be in sewage works. Although improvements can still be made to the existing process, the next new step for this type of treatment could be digestion of the organic portion of domestic refuse. There have already been some experiments on digesting the entire (organic) waste products of human living, but it is obvious that there would have to be some drastic modifications in the process to do this effectively.

However, it is in the treatment of agricultural wastes that anaerobic digestion holds the most immediate promise for those farmers prepared to do some experimentation. What is needed more than anything else in the field of anaerobic digestion is for the process — now that it has been shown to work — to be tried out more extensively in full-scale farming situations. It is only in this way that the problems will come to light and finally be solved — problems such as the collecting of waste and the stirring and heating of the digester.

Nowadays neighbouring farmers often share expensive equipment in order to minimise costs; the idea of 'co-operative dung' is merely an extension of this. If farms are close enough that the cost of transporting the animal waste to the digester is not too large, the production of methane from the combined dung could be worthwhile. If the distances become too great, the total energy balance is disturbed and the system would become a net energy consumer.

Although anaerobic digestion offers a useful method of pollution control, the benefit which is likely to be the most attractive in the long run is the production of methane. As a renewable energy source methane from organic wastes will become more significant than the present national figures for fuel consumption suggest. With a gradual or sudden decline in the quantities of energy used per head — methane will come into its own. The investment in a methane digester at today's prices will mean that the financial return will be that much greater as fuel costs escalate.

To build a digester now is to prepare for the future, for if the refinement of the process is not carried out while other resources are available, it will be too late to do so when they have run out. It is the duty and responsibility of everyone who has the means to show people today that digestion works and to improve it for people tomorrow.

Glossary

Aerobes/aerobic — bacteria which require the presence of free oxygen for their metabolic processes. Oxygen in chemical combination will not support aerobic organisms.

Anaerobes/anaerobic — bacteria that do not require the presence of free or dissolved oxygen for metabolism. Strict anaerobes are hindered or completely blocked by the presence of dissolved oxygen; facultative anaerobes can be active in the presence of dissolved oxygen but do not require its presence. Hence an aerobic system is one which requires oxygen. Hence an anaerobic system is one which does not require oxygen.

Algae — primitive plants, one or many-celled, usually aquatic, and capable of synthesising their own foodstuffs by photosynthesis.

Alkalinity — a quantitative measure of the capacity of liquids to neutralise strong acids. Alkalinity results from the presence of bicarbonates, carbonates, hydroxides, volatile acids and salts.

Activated sludge — a process of waste treatment used to biologically degrade organic matter in a dilute water suspension. High-rate air diffusion through the liquor promotes the growth of aerobic bacteria and other organisms, which, acting on the organic matter, produce a sludge.

B.O.D. (Biological Oxygen Demand) — an indirect measure of the concentration of biologically degradable material present in organic wastes. It is the amount of oxygen used by aerobic bacteria when allowed to degrade the organic matter in the presence of air at a constant temperature (20 °C) for a specific time (5 days). It is expressed in milligrams of oxygen used per litre of liquid waste (mg/l) or in parts per million (p.p.m.).

Biodegradable — organic matter which can be broken down by bacteria or other organisms is said to be biodegradable.

C/N (Carbon/Nitrogen ratio) — the weight ratio of carbon to nitrogen.

141

C.O.D. (Chemical Oxygen Demand) — an indirect measure of the amount of oxygen required to degrade the organic matter in water chemically. It is a measure of both the biodegradable portion and the portion of the waste which can only be oxidised chemically. It is determined by the amount of potassium dichromate consumed in a boiling mixture of chromic and sulphuric acids. The C.O.D. will always be higher than the B.O.D. value.

Calorific value — the energy content of fuels; it represents the amount of heat released when the fuel is completely burnt. It is measured in BTUs/cu. ft. (or per lb) or Joules/cm^3 (or per Kg.).

Eutrophication — the process in which lakes and rivers have an increasing nutrient content, either due to natural run-off from the land, or due to treated effluents from sewage, farm wastes and fertilisers etc. The excess nutrients promote growth of algae, which can block up the river and cause oxygen depletion when they die. Too many algae in the water can also cause problems in filtration at water works.

Ion — when a salt dissolves in water, it dissociates, i.e. splits-up into two charged ions (a positive and a negative), e.g. common salt, dosium chloride, NaCl in water becomes Na^+Cl^- .

Liquefaction — the process of breaking down and dissolving organic solids, by the action of bacterial enzymes outside their cells.

Leachate — water which flows through the soil or ground and ends up in a water course. If the water is contaminated initially and does not receive enough purification as it passes through the soil, the leachate can cause pollution. Similarly if rain etc. falls into a highly organic sludge it can carry polluting material with it.

Methanogenesis — the process of producing methane, hence

Methanogens — the bacteria which produce methane.

Pathogen/pathogenic — bacteria which cause disease both in animals and humans, e.g. Salmonellae cause typhoid.

pH — the measure of acidity or alkalinity. It is the logarithm of the reciprocal of the hydrogen ion concentration. pH7 is neutral, less than 7 is acid, more than 7 is alkaline.

Oxidise/oxidation — the addition of oxygen to a molecule. In the case of organic matter the molecule is often broken down to a smaller one. In aerobic organisms the end product of oxidation of organic matter is carbon dioxide and water. When methane is burnt it is completely oxidised to CO_2 and H_2O. The removal of hydrogen from a molecule is also a form of oxidation and thus the formation of methane is the methane-bacteria's way of obtaining oxidising power.

Percolating filter — a form of aerobic waste treatment, involving the distribution of the water and dissolved organic matter over 'beds' of stones, clinker, plastic media etc. upon which the aerobes grow. As the organic waste trickles past them, they absorb it and break it down. As they grow, they gradually fall off the media and flow out with the effluent as a sludge. This is then settled out and the treated water can be discharged to the river etc.

Rumen/ruminants — herbivores, such as cows and sheep, have a multiple-chambered stomach, the rumen, which has evolved so that they can digest the grass and cellulose that they eat. Inside the rumen live anaerobic bacteria (amongst them methanogens) which break down the cellulose to smaller carbohydrate molecules which the ruminant can absorb.

Septic tank — a single-story settling tank in which the organic portion of settled sludge is allowed to decompose anaerobically without removal or separation from the bulk of the water flowing through the tank. Only partial liquefaction and gasification of the organic matter is accomplished and eventually the undecomposed solids will accumulate to the extent that solids removal is necessary.

Supernatant — the liquid which collects on the top of a waste or sludge after the solids are allowed to settle. By removal of the supernatant the water plus dissolved organic matter can be separated from the solids.

Suspended solids — solids that float or are in suspension in water and which can be removed by filtering.

Total solids — the residue remaining when the water is evaporated away from a waste and dried at $110°C$.

Volatile solids — the portion of the total solids which is driven off as volatile gases when the sample is heated to $600 °C$ for one hour. Often synonymous with organic matter content.

N.B.   Some of these definitions were taken and adapted from the MAFF Technical Report No. 23 Farm Waste Disposal, K. Jones, 1970.

Bibliography

This bibliography is obviously not comprehensive, but will provide a basis for further reading. More references can be found in both of L.J. Fry's books and in 'Methane: Fuel of the Future', and in due course from the various more complete bibliographies at present being compiled.

1. American Public Health Association. (1969). Standard Methods for the Examination of Water and Waste Water. 12ed. New York.
2. American Chemical Society. Advances in Chemistry Series No. 105. Anaerobic Biological Treatment. (1970).
3. Bell, Boulter, Dunlop and Keiller. Methane: Fuel of the Future. Andrew Singer, Cambridge. (1973). Prism Press. (1975).
4. L.B. Escritt. Sewerage and Sewage Treatment. C.R.Books London. (1965).
5. L.B. Escritt. Sewers and Sewage Works. Allen & Unwin London. (1971).
6. Eckenfelder and O'Connor. Biological Waste Treatment. Pergamon Press. (1961).
7. L.J. Fry. Practical Building of Methane Power Plants. D.A. Knox, Andover, Hants. (1974).
8. L.J. Fry. Methane Digesters for Fuel Gas and Fertilisers. (1973).
9. Imhoff, Müller and Thistlethwaite. Disposal of Sewage and other water-borne wastes. Butterworths. (1966).
10. Institute of Civil Engineers and Ministry of Housing. Safety in Sewers and Sewage Works. London. (1969).
11. L. Klein. River Pollution 3 Vols. Butterworths. London. (1966).
12. McKinney. Microbiology for Sanitary Engineers. McGraw Hill. (1962).
13. J.C.V. Mitchell. Biogas Today — a producer's handbook. (1975).
14. J. Priestley. Industrial Gas Heating. Ernest Benn Ltd. London. (1973).

15. S. Sampson, ed. A McKillop. Methane Atomic Rooster's Here. Low Impact Technology. (1974).

Journals and Periodicals
1. Appropriate Technology (ITDG)
2. Critical Reviews in Environmental Control
3. Compost Science
4. Effluent and Water Treatment Journal
5. Environmental Pollution Management
6. Farmers' Weekly
7. Fertiliser News
8. Journal of Water Pollution Control Federation
9. Journal of Institute of Water Pollution Control (previously Journal of Institute of Sewage Purification)
10. Process Biochemistry
11. Soil Science
12. Water Research
13. Water and Waste Treatment

Useful equations and Conversion Factors

1. Areas of a rectangle = Length x width
   Areas of a circle $= \pi \times (\text{radius})^2$

   Surface area of a cylinder
   $$= 2\pi \times \text{radius} \times \text{height} + 2\pi \times (\text{radius})^2$$
   Surface area of a sphere $= 4\pi (\text{radius})^2$

2. Volumes of a rectangular tank
   $$= \text{Length} \times \text{Width} \times \text{Height}$$
   Volumes of a cylinder $= \pi \times (\text{radius})^2 \times \text{Height}$
   Volumes of a sphere $= \frac{4}{3}\pi \times (\text{radius})^3$

3. Retention time $= \dfrac{\text{Volume of Tank}}{\text{Rate of flow of liquid or gas}}$

4. 1 inch = 2.54 cms.   1 m. = 3.28 feet
   1 foot = 0.305 m.   1 sq. ft. = 0.093 sq. m.
   1 sq. m. = 10.76 sq. ft.   1 cu. ft. = 28 litres = 0.028 m^3.

5. 1 gallon = 4.55 litres   1 litre = 0.22 gallons
   1 m^3 = 220 gallons

6. 1 gallon occupies 0.161 cu. ft.   1,000 litres in 1 cu. m.
   1 cu. ft. = 6.23 gallons

7. 1 gallon of water weighs 10 lbs.
   1 litre of water weighs 1 kg.

8. 1 lb. = 0.454 kg.   1 kg. = 0.221 lb.
   2240 lb. = 1 ton = 1.02 tonnes   1000 kg. = 1 tonne = 0.984 tonne

9. 1 part per million = 1 milligram/litre = 1 gm/m^3
   1% = 10,000 ppm = 10,000 mgm/litre = 10 gm/litre

10. 1 cu. ft./lb. = 0.062 m^3/kg.   1 m^3/kg. = 16.1 cu. ft./lb.

11. 1 lb./cu. ft. = 16.2 kg./m^3.   1 kg./m^3 . = 0.062 lb./cu. ft.

12. 1 acre = 4,840 sq. yards = 0.405 hectares
    1 hectare = 10,000 sq. metres = 2.47 acres

13. $T°F = \frac{5}{9} (T - 32)°C$ $T°C = \frac{9}{5} T + 32°F$

14. Pressure: 1 in. of water = 0.25 m. bar
               1 lb/sq. in. (psi) = 68.95 m.bar
               1 atmosphere = 1.013 m.bar

15. 1 British Thermal Unit = 0.252 Kcals.
    1 BTU = 1,055 Joules   1 Kcal = 3.97 BTU
    1 Joule = 9.5 x 10^{-4} BTU   1 Kcal = 4.19 kJ.

16. 100,000 BTU = 1 Therm = 29.3 kilowatt hours

17. 1 BTU/cu. ft. = 0.038 J/cm^3 (or Mega Joules/m^3)
    1 J/cm^3 = 27.0 BTU/cu. ft.

18. 1 BTU/lb. = 2320 J/kg.   1 J/kg. = 4.29 x 10^{-3} BTU/lb.

19. 1 BTU/hr. = 0.0011 mJ/hr.

20. Heat transfer coefficient:
    1 BTU/ft.^{2} /° F/hr. = 20.44 kJ/m^2 /° C/hr.

Index

Acetic Acid (acetate). 7, 18.
Acidity. 13-15, 18, 49.
Acid Formation. 9, 10, 14.
Acitivated Sludge. 2, 22.
Acts — Control of Pollution 1974.
  — 22, 110, 111.
  — Health and Safety at Work.
    1974. 111.
  — Industry 1972. 112.
  — Public Health 1936.
    23, 110.
  — Rivers (Prevention of
    Pollution) 1951 and
    1961. 110.
  — Water Resources 1963. 110
Advisory bodies. 136.
Aerobic bacteria. 7, 9.
    treatment. 22.
Aga cooker. 73.
Agricultural Development Advisory
  Service. 136.
Agricultural wastes. 21, 111.
Air leaks. 61.
Algae. 27, 84
Ames, Crosta, Mills Ltd. 130.
Ammonia/ammonium. 10, 12, 13,
  18, 25, 93.
Anaerobic bacteria. 7, 9.
    contact. 43-46.
    filter. 43-46.
    treatment. 22, 23, 24, 27.
Ashwell. 124.
Auchincruive (West of Scotland
  College). 94, 116.
Autonomous House. 78.

Batch Digester. 35, 36, 47.
Bedding material. 87.
Biogas composition. 28.
    Volume/weight of organic
     matter. 30.
    Calorific value. 30.
    Inflammability. 31.

Biogas Plant Ltd. 36-38, 53, 63,
  103, 124-126, 135.
Biological Oxygen Demand. 21.
Biomechanics Ltd. 48, 94, 126,
  127, 135.
British Standards Institute. 112.
Building Regulations. 110.
Building Research Establishment.
  112.
Burner. 74, 75.

Calorific value. 30.
Cameron. D. 2.
Carbon/Nitrogen ratio. 12, 35,
  92, 93, 98.
Carbon content. 12.
Carbohydrates. 6, 7, 10, 29.
Cars. 76, 77.
Cellulose. 6, 12.
Chemical Oxygen Demand. 21.
Cherry Valley Duck Farm. 123.
Compression of gas. 77.
Concentration of sludge. 95.
Condensate traps. 58, 69.
Conservation Tools and Techn-
  nology Ltd. 134, 135.
Cooking. 72.
Cost-benefit analysis. 113-115.
Coulthard, J. 77, 133, 134.

Department of Energy. 112.
Developing countries. 3, 79.
Digester, design check list. 105.
    dimensions. 102-103.
    shape. 51, 98, 102-103.
    sizing. 93.
    site. 104.
Dilution. 96.
Displacement digester. 36, 43, 47,
  103.
Dorr Oliver Ltd. 63, 130. -

EIMCO Ltd. 130.

148